Building Surveys

Third edition

Peter Glover
FRICS

An imprint of Butterworth-Heinemann

Laxton's
An imprint of Butterworth-Heinemann
Linacre House, Jordan Hill, Oxford OX2 8DP
225 Wildwood Avenue, Woburn, MA 01801-2041
A division of Reed Educational and Professional Publishing Ltd

℞ A member of the Reed Elsevier plc group

OXFORD   BOSTON   JOHANNESBURG
MELBOURNE   NEW DELHI   SINGAPORE

First published 1983, as *Surveying Buildings*
Second edition 1990, as *Building Surveys*
Third edition 1996
Reprinted 1998

British Library Cataloguing in Publication Data
A catalogue record for this title is available from the British Library

Library of Congress Cataloguing in Publication Data
A catalogue record for this title is available from the Library of Congress

ISBN 0 7506 3218 6

Printed and bound in Great Britain by MPG Books, Bodmin, Cornwall

Preface

The first edition of Building Surveys was published in 1983, a readable volume, essentially factual and practical rather than controversial or opinionated. In the preface to the first edition the hope was expressed that surveyors might find the contents helpful to them in their day to day work and that students might find the book instructive.

Reaction was very favourable indeed and a second enlarged edition was published in 1990. Since then there have been further developments which warrant an updating of the text. Litigation remains a problem with the cost of professional indemnity insurance continuing to rise in response to the number and value of claims against surveyors. There have been changes in legislation, particularly Building Regulations. This third edition takes these matters into account and reflects where appropriate the results of some of the more recent court cases in which surveyors have been unfortunate to feature, particularly 'Watts v. Morrow' where the Judge was asked to consider the position of a surveyor who departed from the RICS guidelines and dictated his report directly on site with no retained file notes.

The advice given has also been reviewed where necessary to take into account the views and opinions expressed by practising surveyors who have been kind enough to comment on the previous editions.

Acknowledgements

This book is dedicated to: all those surveyors who have contributed over the years to our knowledge of building performance and building defects, passing on their wisdom and experience in the process; to the clients who commission Building Surveys and ultimately ensure the continuation of the service; to Nationwide Building Society who gave permission for the reproduction of their Mortgage Valuation Report Form; Rentokil Laboratories who provided illustrations of selected wood borers and timber defects; and Nelson Hurst and Marsh Ltd, Insurance Brokers specialising in surveyors' professional indemnity insurance, who provided the relevant sections in Chapter 19.

Contents

Frontispiece Equipment for a building survey of a simple low-rise structure

Introduction

If the prosperity of a nation is measured by the value of its fixed assets and if the optimism and energy of its people is reflected in the way in which they maintain and improve their buildings, then it is clearly of the greatest importance that a country's building stock is kept in good repair. It is important that those involved in buying or leasing buildings, or lending money on the security of property, are well advised. This book is written as a practical guide for surveyors and others who may be asked to survey buildings and prepare reports. Although not primarily intended as a text book on building construction, it is also aimed at the student who, whilst perhaps familiar with basic building construction, lacks the knowledge or experience of building performance.

The preparation of reports on buildings is a very personal service and since no two surveyors and no two clients will be the same, the style and content of reports can vary considerably from case to case, from practice to practice and in different localities. The Royal Institution of Chartered Surveyors and the Incorporated Society of Values and Auctioneers have jointly published Guidance Notes for residential Building Surveys. The intention of these Guidance Notes is to introduce some uniformity of procedure and improve the general standard of Building Survey Reports. Great stress is placed on the need for surveyors to confirm the nature of their instructions before undertaking this work, in order to avoid misunderstanding and disputes with clients.

At a time when the professions are open to searching criticism and the public are becoming more litigation conscious, it is not surprising that claims against the professions from their clients are increasing in numbers and value. All professions are affected by this trend but there are few areas of activity where the potential for significant error giving rise to serious financial losses to the client can be as great as in the field of building surveys, especially where the larger type of commercial and industrial structures are concerned.

Alongside the tendency for surveyors' reports to become longer and for clients to expect more from their surveyor there has been an increasing use of abbreviated report forms. Building Societies, Banks and other mortgage lenders routinely send their customers copies of valuation reports and despite warnings to the contrary there is no doubt that many of these customers rely on such documents and fail to obtain their own independent report. In addition various pro-forma style reports are on offer variously termed Housebuyers' Reports, Homebuyer Reports or Flatbuyer Reports. In response to this trend, a chapter on the use of such reports has been included in this book with some guidance as to the procedures to be followed. Later chapters deal with commercial and industrial buildings, flats, new and older buildings, and leasehold property reports as well as reports for mortgage purposes.

Publications mentioned in the text are listed in the bibliography so that the source of additional information on a particular topic may readily be found.

The specimen report in Chapter 18 has been included only for guidance as to the layout and wording which may be adopted in typical reports. It contains some useful standardised wording. The intention of using standard report clauses is that they can make report drafting easier, whether or not a word processor is used, in that they allow the rapid drafting or dictation of those parts which are routine, permitting the surveyor more time to devote to the particular and specific problems of the building under review.

Obviously all possible steps should be taken to avoid wasting time on routine report drafting or dictation from notes since the surveyor's time is his most valuble resource and it must be used efficiently.

Finally, a comprehensive index has been provided with references to the sections concerned in the book and sources of additional information where known.

The nomenclature used in the book and in the reports is that which is the most simple and straightforward. In respect of the reports which the reader may prepare, the writer would recommend the use of simple straightforward English. Technical expressions should be avoided altogether when writing to a client unless they are known to be understood by that client or are explained at the time. Similarly, the writer has used spellings which the clients would expect rather than those which some surveyors may prefer. For example, throughout the text 'sill' is used rather than the more traditional surveyors' 'cill', and I have also used 'lintel' as being more suitable than 'lintol' for the same reason. Clients seeing 'cill' and 'lintol' in a report assume that a spelling mistake has been made, thus making the accuracy of the report suspect in their minds.

During the course of preparing the book, a selection of files for properties inspected and reported upon over the last thirty years was studied. Considering the nature of the properties involved and the variety of clients, I had cause to reflect on what a very mixed bunch they were. One point standing out from this evidence was that despite the almost inexhaustible range of possible building defects which could arise, a great deal of building construction is still basically sound. Reports which dealt with minor defects in a routine manner were the rule rather than the exception, most buildings being structurally sound. In only about 5% of cases were the buildings in such poor repair or subject to such serious defects that the client was advised not to buy or lease them unless prepared for substantial ependiture to put things right.

The main question from clients was, 'Is this price (or rent) fair and reasonable?'. In order to answer this question, surveyors need knowledge of both building construction and property values in the locality because, at the end of the day, after everything has been said about the leaking roof, the damp, the furniture beetle and a multitude of other potential hazards, it is that simple question which has to be answered.

P G

Chapter 1

Surveyors and their instructions

It is highly important to ordinary members of the lay public that a surveyor should use proper care to warn them regarding matters about which they should be warned over the construction or otherwise of a piece of property and that they should be told what are the facts.
Mr. Justice Hilberry – Rona v Pearce (1953)

The Oxford Dictionary defines a surveyor as, among other things, 'a person professionally engaged in surveying', and to survey as 'let the eyes pass over, take general view of, form general idea of the arrangement and chief features of, make or present survey of; examine condition of (building etc); determine boundaries, size, position, shape, contour, etc, of (land, coast, district, estate, etc);' whence surveying!

The primary purpose of a building survey is to give an independent professional opinion on the structural condition of a property, but the word 'independent' as used here probably requires further explanation. Strictly speaking, the surveyor carrying out the building survey cannot be wholly independent since he or she is acting on behalf of a client, the purchaser, whose interests he or she is serving. However, the main object of such a survey is to ensure that an opinion is given by someone who has no connection with the vendor or his agents. In the same vein, the surveyor acting for a building society cannot be regarded as wholly independent since of necessity he or she must consider the building society's interests before those of a prospective purchaser, although the surveyor will have some liability to the purchaser for the advice given if it is known that a purchaser is likely to rely upon it (see Chapter 19). As far as the description 'professional' is concerned, those who are best qualified to carry out such a survey must have had considerable experience in the performance of buildings and their components as well as their design and construction.

The survey

Since the end of the Second World War, the expression 'structural survey' has entered the vocabulary of those engaged in the property market and their advisers. This term, not ideal, has been generally used throughout the United Kingdom, and surveyors, estate agents, solicitors and bank managers, as well as many members of the public engaged in buying and selling houses have become familiar with it and the nature of the service it embraces. In the writer's view the term 'building survey' is more suitable and has been recently adopted in place of 'structural survey' by the surveying profession as a better description of the service.

A building survey report provided by a surveyor when a house is likely to change hands is now a commonplace occurrence. While many houses are still sold without the benefit of such a report, in an inceasing proportion of cases a building survey *is* involved and evidence suggests that about 10% of house buyers now commission their own report.

It is interesting to reflect that the provision of this service to the general public is a recent development brought about by a rapid increase in house ownership since the Second World War and an increasing awareness among the public of the possible pitfalls involved in owning defective property or buildings needing expensive upkeep. The present-day consumer expects value for money, is more knowledgeable in respect of the building he or she occupies, and has standards higher than those of previous generations.

The surveying profession may fulfill two roles in undertaking building surveys of houses. First, clients can be advised and, if necessary, they can also be warned. Sometimes a great deal of expense and heart-ache may be avoided for a client who secures a surveyor's report on a building suffering from important defects which, although not apparent to the untrained eye, are obvious to the surveyor. A second role which surveyors may rightly assume is that of front-line professionals who daily come into contact with the nation's building stock and by careful examination and considered advice, may assist in the better maintenance of that stock for the benefit of the community as a whole.

It is also interesting and salutary for the practising surveyor to consider procedures followed in other countries. The house-buying public in the United Kingdom is fortunate in having available a service which is unrivalled elsewhere. In many parts of the world, even in highly-developed countries, the provision of professionally-prepared reports of the type available in the UK is unknown. This is an important service which must be encouraged and, where necessary, improved, and competent, well-trained surveyors are essential for the maintenance of this service in the future.

The surveyor

Before undertaking building surveys, the surveyor should be qualified both academically and by practical experience to a high standard, but there is at present no statutory requirement for this to be so. Anyone may set himself (or herself) up as a surveyor and accept instructions from the general public to underake such surveys. Undesirable as this may seem, there is no obvious prospect of the situation changing in the foreseeable future, although an unqualified surveyor without a professional indemnity insurance policy is taking a very high risk. On the issue of academic qualification it must, however, be said that there are many members of the profession who have had little or no examination success in the formal sense but who are nevertheless well qualified by their long experience and detailed knowledge to prepare reports to the highest standard.

Similarly, there are younger professional members who embark upon their careers lacking practical knowledge, having only examination success to start them off. The decision as to whether or not a newly-qualified surveyor is ready to undertake this complex and difficult work is not easy to take. Preferably, an older, more experienced surveyor should be available to advise his younger colleague on the timing of this decision based upon some personal knowledge of that younger man and his professional skills.

In case the reader may be a young man or woman contemplating surveying as a career, or an older man or woman advising on career choice, the writer would mention briefly the features which would seem to him to be needed by an individual wishing to undertake this work. There are, of course, other branches of the surveying profession to which these comments may not necessarily apply.

First, a surveyor should have good command of the English language, the ability to express himself or herself in reports with clarity and the ability to express himself or herself to his clients with verbal conviction.

Secondly, the surveyor should have a practical turn of mind. He or she should be capable, to a reasonable extent, of understanding the processes of building construction and the working of services such as heating systems, and should be able to assimilate new ideas and building processes. It is helpful, although not essential, to be able to undertake simple building operations oneself since in many situations a personal knowledge of basic carpentry, brickwork operations, concreting, electrical wiring and plumbing can be of inestimable value.

Thirdly, the surveyor should acquire a tactful and pleasant personality. When inspecting houses before purchase it will be necessary to visit a vendor's home in circumstances which may be stressful for the vendor or his wife and to spend a considerable time there with the intention of finding faults. One should never underestimate the anxious anticipation which will precede the visit and the fact that the vendor assumes that each note taken indicates a defect found or that from his point of view the longer the time spent on the inspection, the worse will be the resulting report. The surveyor should explain what he or she is going to do and should first ask for permission to proceed.

The foregoing may seem trivial or even obvious to the reader, but it is surprising how often, in practice, one hears of surveyors who have been less than courteous, left unrepaired damage or engendered in the vendor a general feeling of hostility. Surveyors are ambassadors of their profession and their clients and should never lose sight of this fact.

Equipment

The surveyor requires certain items of equipment for use during the survey, but what is considered necessary tends to vary from one firm to another. The following are items which the writer has always found helpful.

First, lighting is needed for roof spaces, under floors and other dark areas in order to ensure a thorough inspection of timbers and structure. It is recommended that a protected lamp-holder be carried with an extension lead so that mains-powered lighting can be used whenever mains electricity is available. There is no real substitute for mains-powered lighting for this purpose and a lead-light with 100 to 150-W bulb should be carried. A selection of electric socket adaptors can also prove useful.

Some lighting should also be carried for those situations where no mains electricity is provided. Powerful torches with spare batteries and bulbs are used by the writer. Lead-lights which run off 12-V car batteries are also used and may have advantages.

Tapes for taking dimensions are necessary. The writer uses short steel tapes for taking dimensions in small rooms and of timber sizes, also a longer linen tape for longer distances. A folding wooden rule is useful for some purposes, for example it can be inserted between boards in a timber floor to check the depth of section of the floor joists.

Ladders are essential for gaining access to roof voids and external roofs. Lightweight aluminium ladders are recommended for this purpose and the writer uses aluminium 'surveyors ladders' which dismantle into four sections for fitting into a car boot.

Important attributes for any surveyor are that he or she has a reasonable head for heights and is sure-footed when engaged on inspections above floor level. A reasonable standard of fitness is also required since there will be many occasions when the surveyor, rather than an assistant, must look closely at a particular item. It may not be possible to stay at ground level and rely upon second-hand details taken by a helper. The writer as a student was sent up a long extension ladder to the roof of a block of flats, not, as he discovered later, because his valued opinions were required but in order to test his resistance to vertigo. Access to roof voids, the roofs themselves and underfloor areas often require strenuous acrobatic feats.

Any surveyor should have available at least one good moisture meter for testing woodwork, plasterwork and other areas for dampness. The meter dial should be checked by shorting out the probes on each occasion before the instrument is used, and it should also be zeroed to ensure that it is functioning correctly and that the batteries are not expended. It is important that an instrument with failing or flat batteries should not be used since its readings will then be inaccurate and unreliable.

Various tools and other useful items which can be carried in the car boot include a pair of wellington boots, a spade for taking subsoil samples, assorted drain plugs, and manhole cover-lifting keys, together with a hammer and cold chisel for tight covers. Mirrors, too, are very useful. A small mirror can be used to look along underground drains between manholes and to check for tree-root intrusion, stepped fractures in the pipework and other obstructions. Mirrors are also invaluable for providing a wide view under floors and in other areas difficult of access. A good quality long spirit level for checking walls and floors is invaluable, as is a plumb-line. A pair of binoculars should be used for gaining a clearer view from ground level of roof details and upper storeys.

For general purposes the following equipment is recommended:

(1) Folding or sectional loft ladder.
(2) Linen tape.
(3) Moisture meter.
(4) Mains voltage protected insulated light and extension lead.
(5) Torch.
(6) Clip board.
(7) Claw hammer.
(8) Binoculars.

(9) Long and short spirit levels.
(10) Plumb line.
(11) Steel tape.
(12) Notebook.
(13) Hand mirror.
(14) Spade.
(15) Manhole keys.
(16) Drain plugs.
(17) Folding wooden rule.
(18) Wood probe.
(19) Protective gloves and face mask.
(20) Endoscope.
(21) Camera.
(22) Floor board cutting saw.
(23) Magnetic compass.
(24) Wellington boots.
(25) Assorted screwdrivers, tools and electric socket adaptors.
(26) Identity documents.

More specialised inspections may require the following:

(1) Concrete cover meter for checking thickness of concrete cover to reinforcement bars.
(2) Capacitance moisture meter for detecting dampness behind panelling etc.
(3) Carbide moisture meter for accurate damp detection.
(4) O.S. and Geological Survey Maps.
(5) Metal detector and stud sensor for tracing concealed metal services and locating timber studs.
(6) Concrete reinforcement corrosion detector for pinpointing defective reinforcement.
(7) Inflatable bags and smoke testing equipment for drains tests.
(8) Coloured dyes for tracing drain runs.
(9) Concrete core sampling equipment.
(10) Dumpy level or water level for checking levels in larger structures.
(11) Sonic tape for facilitating single-handed measurement in areas of difficult access.
(12) Digital level meter for measuring angles and gradients.
(13) Schmidt hammer for measuring compressive strength of bricks and blocks.
(14) Hand auger for taking soil samples.
(15) Shear vane tester for measuring shear strength in clay soils.
(16) Calibrated tell tales or studs for monitoring crack movement in walls.
(17) Dew point indicator to detect condensation.
(18) Hand thermograph for measuring heat loss in external walls.
(19) Ultrasonic leak detector for detecting small leaks in pipes from pinholes or fractures.
(20) Radon detector – measuring device for assessing quantity of radon gas in granite formation areas.
(21) Laser measure for facilitating single-handed measurement of inaccessible areas.

Instructions

A surveyor's instructions may be limited to inspecting a particular building and providing a report without any need to comment on the price or related matters, so that the report will be purely concerned with the construction and condition of the property; such reports were (as already discussed) generally termed 'Structural Surveys' although in the writer's view they might better be described as 'Building Surveys'. In many cases the report will also include advice as to value when it might be termed a 'Building survey and valuation'. Most pro-forma reports for banks and building societies are essentially valuation reports which include some superficial comment on condition, and the RICS 'House-Buyer's Report and Valuation' includes a valuation automatically as part of the service. In many cases therefore the question of value is fundamental to the advice being given.

In order to be able to advise in this way, the surveyor should be familiar with property values in the locality concerned. It can be dangerous for a surveyor to accept instructions to value property either in an unfamiliar locality or of a type with which he or she is not normally professionally involved. There will be exceptional cases, of course, where a client will ask a specific surveyor, in whose judgement he trusts, to travel some distance and report on a property, and it would seem quite reasonable for such instructions to be accepted provided that the client understands the implications and the possible limitations, points which should be confirmed in the surveyor's letter of acceptance of the instructions.

Again, there are many specialised types of property such as hotels, licensed premises, garages and others where special markets exist and where specialist firms of surveyors might best be consulted on questions of value. Where the building construction itself and the subsoils may be unfamiliar, it would not seem unreasonable for a surveyor to accept instructions provided that the position is explained and the fee is adequate to cover the additional research which will be necessary.

The RICS Practice Note 'Structural Surveys of Residential Property' recommends that before taking instructions the surveyor discusses the client's requirements and advises on the type of survey needed. In some cases time should be spent with the client assessing his needs. It is considered essential, in the interests of both client and surveyor, that when instructions are given, the surveyor should write to the client to acknowledge their receipt and to confirm their details. In respect of building surveys of houses, the RICS recommended that if possible this be achieved by a covering letter and a short pre-printed booklet describing in detail the extent and limitations of the proposed survey, a system already adopted by some larger surveying practices. At this stage the surveyor should also agree either the fee or the method of assessing the fee (i.e. if an hourly or other rate for professional time is to be charged).

There may be occasions where verbal instructions can be accepted from regular clients already familiar with the procedure and this will normally apply in the case of pro-forma reports for institutional or company clients who will already understand what is involved. The RICS recommendations need apply only to surveys or valuations for members of the public with whom the surveyor is dealing for the first time.

The Practice Note recommends that with respect to surveys of residential property the following matters be confirmed in writing by the surveyor to the client:

(1) The nature of the instructions. If something less than a building survey is to be carried out it is important that the extent of the obligations being assumed by the surveyor are understood and agreed to by the client. Many surveyors decline to take instructions for anything less than a full survey for an intending purchaser, save in most unusual circumstances.

However, there will be many instances where a partial survey of some specific aspect of a building may be undertaken for an owner occupying his own property, perhaps concerned about some specific defect in, say, a roof or a particular floor. Where such a report is prepared for an owner, the letter confirming instructions will make clear the limitations of the advice given.

(2) When the survey is likely to be carried out. Contact should first be made through the vendor or his agent with the occupier of the premises to be surveyed and arrangements made for the inspection to be carried out at a convenient time and date. The occupier should be informed of what is involved in the survey and the extent to which he will probably be inconvenienced. He must be told how long the survey might take and how many people are likely to be present, particularly if specialists are involved in testing and inspecting the services.

(3) The surveyor's intentions regarding:
   (a) The extent of the inspection of the main structure.
   (b) The extent of the inspection of outbuildings and grounds.
   (c) The amount of structural exposure to be carried out.
   (d) The limitations of inspection where surfaces are covered (e.g. by fitted carpets).
   (e) The extent to which furniture will have to be moved.
   (f) Any limitation of liability in the form of a general exclusion clause which will appear in the subsequent report.
   (g) The extent of enquiries to be made of local and statutory authorities.
   (h) The extent to which the survey includes the testing of services.
   (i) The fee payable and method of assessment.
   (j) Whether or not valuations are included. Estimates of reinstatement costs and the costs of recommended works and any limitations and reservations applicable to such works.

In most cases similar points will need to be confirmed when accepting instructions to prepare reports on commercial and industrial premises since the liability is the same and the scope for structural exposure may be more limited due to the nature of the occupation of the premises.

The Practice Note also recommends that appointments made with the vendor should be confirmed in writing with a note of the length of time likely to be involved and the extent to which the surveyor proposes to lift floor coverings, expose the structure or test services. This will not always be necessary. Frequently the surveyor will have to inspect empty buildings to which access will be gained by keys borrowed from estate agents. A letter to the vendor in these circumstances would be inappropriate. If no special exposure work or tests of services are to be undertaken then the survey of an empty building may proceed without any written exchange with the vendor or his agents but otherwise prior consent for opening up or testing should be obtained.

It is recommended that written notes be taken on A4 sheets using a clipboard. A standard 'field sheet' including a checklist of prompts to remind the surveyor of matters to be covered is helpful. Discretion in note taking is advised and discussion of any findings with the vendor best avoided. In view of the possibility of litigation at a later date it is considered essential that reasonably comprehensive site notes are available on file, so direct on-site dictation with no note taking is not recommended.

Organisation

The surveyor's practice requires office premises and staff to be organised so as to provide all necessary assistance to surveyors in the field. Such organisation varies considerably from practice to practice, which range from small one-man firms to major multi-partner practices needing specialist management. Very often the one-man firms can offer a better service since a building survey is a highly personal matter between client and surveyor which, in any event, can be carried out effectively by only one person.

It is suggested that the surveyors' practice should have a library kept up-to-date with all the relevant technical publications as they are published. The Building Research Establishment (BRE) publishes quarterly its *BRE News* as well as digests, information papers and technical papers.

The *Registered House-Builders' Handbook* is essential where new housing is to be reported upon. The writer has found it helpful to retain sections dealing with the previous types of certificate which were issued by the National House-Building Council (NHBC) prior to the current type, since procedures under previous certificates are different from those currently applicable. Older technical requirements where these have been superseded should be retained in order to be able to comment upon houses which are not now new but which are still subject to NHBC protection. Although basic NHBC requirements have not changed, a number of important detailed practice notes have been added to deal with certain specific problems.

A set of Ordnance Survey maps should be kept covering the district in which the practice operates. These can be invaluable by indicating old quarries, mine shafts, gravel workings and suchlike features as well as by providing information on local watercourses which can sometimes indicate where underground water may be present. At the base of a valley, the contours could indicate whether or not a constricted flood plain exists which in exceptional circumstances might pose a flood risk. The geological survey maps also indicate subsoil types and possible areas of fill material where gravel or other workings may exist, and a set of such maps covering the practice area is recommended.

Bomb maps are also available showing where known explosive devices fell in some localities during the Second World War. Occasionally important evidence of structural bomb damage may be confirmed from such maps. Information on bomb damage may also be obtained from the Imperial War Museum. The records are not complete but the author has found them to be helpful on occasions.

Having completed the physical side of the survey, the surveyor returns to the office to draft the report. Whether a word processor is used or not, this should be done using a standard procedure, with a standard set of headings and layout. Some of the report may be compiled using specimen clauses to permit more rapid dictation or drafting and to eliminate as much as possible of the time involved in the more routine parts of the report. The surveyor may then devote time to those parts of the report which are individual and most important to the specific client.

A suitable atmosphere is needed when a report has to be drafted and it is important to be able to work free from interruption and distraction with all the information needed readily to hand. It is also recommended, wherever possible, that each individual deals with one report at a time and completes the drafting of that report before commencing another. No matter how comprehensive the notes, there will often be points relevant to the report which arise from memory and it is clearly essential that the property under review should not become confused with another. Sketches and instant photographs taken on site can be spread out on the desk as an aid to reflective thought.

Clients

Having considered the means by which the surveyor takes instructions and undertakes to provide reports, it is necessary to consider the clients for whom he or she will be acting. The relationship between the surveyor and the client is a contractual relationship on the basis of which a service is provided and payment for that service is made. A relationship also exists between surveyor and client and may exist between the surveyor and third parties under the law of tort.

The report may be read and relied upon by third parties and if the surveyor wishes to avoid third party liability under the law of tort, a specific disclaimer to this effect should be inserted in that report. A particular danger arises when a report and valuation are being used by the client to raise a loan on the security in circumstances where that valuation may have been provided for some other purpose or where the surveyor should have included other or different advice had he known the use to which the report would be put. Cases have even been known of current market valuations being used by ill-advised clients for fire insurance purposes. Another hazard is that a report for a property owner which is related only to some particular aspect of construction, and is therefore not a complete report, may be used by that owner to further the sale of the property at a later date, and would perhaps be relied upon by a subsequent purchaser. In the case of commercial or industrial premises, a specific client running a particular type of business could obtain a report which could later be used by the proprietor of quite a different type of business for whom additional or different advice ought to have been necessary.

As a general rule it is recommended that a specific disclaimer against any liability to third parties be inserted prominently in all reports, although this may not of itself provide complete protection against third party claims. (See chapter 19.)

In the case of clients requiring reports on residential property, I have encountered a number of unusual cases where special treatment was called for. One such case is that of the disabled client who may be unable physically to make a proper inspection himself and who has particular needs for specific doorway widths and ramps for wheelchairs. It is advisable to spend some time ascertaining the specific needs of such a client and to cover those needs in the report.

Reports for blind clients have been prepared on the basis that the report would be read out to them and it is then necessary to reconcile two conflicting needs. A blind client needs a reasonably brief report which is very much to the point since it is otherwise most difficult to assimilate all the information when the report is read aloud. He also needs some description of all those matters which the sighted client will already know about and I have found that a considerable challenge arises in providing such a service.

Elderly clients might be warned about design which may be unsympathetic to their needs and a point should be made of closely checking staircase balustrading and handrails, also tight staircase winders which, although not now permitted under the Building Regulations, may be common in older buildings. Clients having young children should also be warned about this type of hazard and also the potential danger where there are large sheets of ordinary-quality glass in glazed doorways and patio doors.

On occasion, clients will consider buying a building with a view to a change of use, making alterations or carrying out extensions. They should be warned that planning permission may be needed for the proposal they have in mind and that the granting of planning permission can never be regarded as a foregone conclusion. I can recall a number of cases where the planning officer enthusiastically discussed a proposal which was later rejected by the planning committee. The frequency of this occurrence is roughly equal to the number of cases where planning officers are distinctly gloomy about the prospects for a proposal which is later approved without difficulty.

Passing of plans under Building Regulations may also be necessary and could present problems which should be mentioned in the report.

If permission to undertake alterations to or to change the use of a property is essential to the viability of the purchase, then the client should be advised to postpone any exchange of contracts until the outcome of applications for such changes is known and the certificates are in hand.

Clients buying a freehold building with a view to undertaking alterations may require permission under the terms of restrictive title covenants in addition to local and national statutory requirements. This applies especially to residential developments in the higher price brackets where estate covenants exist which could restrict considerably the alterations or changes of use which may be allowed.

For clients taking an existing or new lease and contemplating undertaking alterations or improvements to the premises, the consent of the freeholder and any intermediate leaseholder will generally be needed. In some cases there are certain steps which a lessee may take to safeguard his position and to secure an entitlement to compensation in respect of business premises. If the surveyor is aware of the proposals the client has in mind, he could provide helpful advice, not only on the physical and financial problems involved in undertaking the works, but also on the legal implications.

When it is known that the property to be surveyed has been altered or extended in the past, or its use has been changed, the client may wish to know whether such works had the necessary local authority approvals and any other consents. If not, he should be informed as to the implications which may arise. In the case of residential property, it is often found that small extensions or internal structural alterations have been undertaken, perhaps in an amateur fashion and without any consent being granted under the Building Regulations. Occasionally a clear breach of regulations may be found and the question arises as to whether or not the local authority may wish to take action. In the vast majority of cases no action would be taken since the local authority are unlikely to know of the breach unless their attention is called to it. Building Regulations enforcement proceedings against a property owner will not normally be taken after a period of a year has elapsed from the commission of the breach save in unusual circumstances.

Clear breaches of Building Regulations which appear in recently completed works should be noted and the details be given to the clients who may then take legal advice. In general most breaches will constitute defects which the surveyor will, in any event, discuss in his report, but occasionally such breaches could be purely technical and acceptable to the client such as, for example, natural lighting or ventilation requirements, balustrades and the like. These are then merely to be pointed out to the client who may decide what, if anything, he wishes to do about them.

The possibility of enforcement proceedings being taken against a property owner under the Town and Country Planning Acts could arise in respect of unauthorised buildings, extensions or changes of use. Generally, buildings completed for four years or more are immune to such proceedings. Changes of use, however, need to have been established before 1 January 1964 to be immune, and if a surveyor is aware of buildings or changes of use which could be of doubtful legality, this may be a matter which the surveyor, the client and the client's solicitor should consider. The surveyor is the professional adviser who will actually spend time at the property concerned, whereas the solicitor will rarely see the property and any relevant information which the surveyor can provide is usually keenly appreciated.

Fees

The basis upon which the surveyor's fees will be calculated should be agreed in advance with the client. The RICS Practice Note states that this is one of a number of essential matters which should be confirmed with the client for building surveys of residential property. It would also be necessary for surveys of other types of property except where the work is being undertaken for a client for whom the surveyor regularly acts on an agreed basis, or where a service is provided for prospective mortgagees or other clients and an agreed scale of charges applies.

During the course of the preparation of this book, the author investigated a limited sample of procedures used by the other surveyors in assessing fees. In some practices the fees for residential surveys are assessed at a rate per room of accommodation and in others a scale proportionate to the proposed purchase price. In each case there was some variation having regard to the age of the property. At most times it was found that the charge was related to the amount of professional time involved in the inspection and the researching for the report, the account then being on an hourly basis.

Having considered the variations which are possible, the author concluded that the fairest means of assessing fees would be on the basis of a charge at an agreed rate per hour of professional time taken, plus disbursements at cost. In taking instructions, it should be possible to explain to the client the approximate amount of professional time which would be involved on a specific survey having regard to the value, the size and age of the structure and the specific instructions of the client.

There are no fixed scales of charges for building surveys. Scale 31 of the scales of professional charges published by the RICS simply states that 'A fee by arrangement according to circumstances' should be charged. In so far as a building survey is being undertaken together with a valuation, then it is suggested that it should be possible for the overall cost to be reduced to the client so that a fee based upon a valuation scale, plus the additional professional time taken to prepare the extra structrual information, might apply. Although some surveyors have compiled their own scale rate of fees for building surveys based on a percentage of the offer price of the property, this can be patently unfair against one party or the other. Take, for instance, the examples of a cottage with a very low offer price and minimal accommodation but in very poor order and requiring a considerable amount of time spent on the survey compared with a large expensive house in exceptionally good condition. The former could quite well prove more expensive in terms of time than the latter.

Increasingly, surveyors will find that they are instructed to prepare a valuation for a prospective mortgagee and at the same time carry out a building survey for the prospective mortgagor. In such circumstances a reduced overall fee should normally apply as compared with the aggregate of the fees which would apply if the matters were dealt with separately. It is recommended that particular care then be taken to establish at the outset exactly who will be liable for which fee and to bill accordingly.

Traditionally, professional persons render accounts after services have been given and often quite a little while later, especially in the case of regular and retained clients. However, having regard to the possibility of bad debts arising, especially when residential property is being surveyed, an account may be rendered with the report in cases where the client was not previously known to the surveyor. Steps should then be taken to recover fees as promptly as possible in such cases. If a client does not buy the house but moves elsewhere, it may be difficult to trace him in order to recover overdue charges.

All surveyors' offices therefore should have a firm procedure for following up outstanding accounts and recovering fees as promptly as possible, and such matters should not be allowed to drift.

In some situations it may be necessary to ask for a payment on account in advance for building surveying or other services, and in the writer's experience and that of the sample of surveyors with whom this was discussed, this is now increasingly common practice.

As a last resort, a client can be sued for non-payment of fees through the County Court procedure for small claims. The amount of fees for most structural surveys on a domestic scale would come within the limits set for County Court claims but it should be emphasised that this procedure should not be contemplated unless it is a 'one-off' client whom you know to be deliberately perverse. In these circumstances, such claims are not normally defended but it is a risk one has to take and although easy to make the application, if the defendant decides to counterclaim this could prove to be a most expensive and time-consuming matter, not to mention the bad publicity involved. Further details may be obtained from a booklet *Small Claims in the County Court* published by HMSO.

Preliminaries

Notes required during the course of the survey may be taken in writing on A4 or similarly-sized paper attached to a clipboard. This enables reasonably-sized sketches to be included. These notes, together with any correspondence relating to the matter and a copy of the final report, may be bound together durably in a permanent file and retained indefinitely in the surveyor's filing system, suitably referenced for identification should this be needed in the future. On site photography is always worthwhile, as may be seen in the example provided in Chapter 18. For the report in Chapter 18 a 35mm film of 24 exposures was used to record various aspects of the survey. Instant cameras are also very useful although the quality of print is not as good.

Having regard to the potential legal liabilities involved in this type of work and that actions in tort might be taken against a negligent surveyor some considerable time later, it is felt that it would be unwise to destroy old files relating to building surveys. Apart from the question of legal liability, which may never arise in a well-run professional practice, there is also the fact that at some time a surveyor may find himself inspecting the same house again for some other purpose, perhaps on re-sale or to prepare design drawings for alterations. An original file in such circumstances could prove invaluable.

One of the well-known features of office life is that immediately old files are destroyed, perhaps having collected dust for many years, invariably something occurs which could have been dealt with better by recourse to the destroyed information. In larger professional practices, shortage of space may require microfilming of important old files so that the information contained may be retained in a more compact yet accessible form.

An office record system should include cross-references to all old files so that these are identified by the address, the client's name and the nature of the subject matter. In the future it would then be possible to look up old reports prepared for properties in the same area or in respect of the same building; also reports prepared for the same client over the years can then be compared and reports of an unusual nature may be compared with similar previous records. For example, in cases where, say, garages, public houses or ecclesiastical buildings are occasionally inspected, reference may be made to files for properties of a similar type in order to compare the approach adopted in the past.

In so far as the office records system deals with property values and building costs, these may also be recorded by reference to property type, size and characteristics, then cross-referenced to the main filing system. Since property valuation is essentially the art of comparing one property with another, accurate records of prices and rents for properties in the surveyor's area are essential if valuation opinions are to be soundly based.

In order to deal adequately with the full range of instructions likely to be received, a surveyor should be able to call upon assistance from other specialists and trades as required. It is helpful if continuing relationships be fostered with specialists and some trades with this in mind. A local firm of structural engineers will be needed from time to time, a jobbing builder with long extension ladders and tools for undertaking exposure work will be necessary to assist with certain types of inspection, and a plumber and electrician will be competent to test the drains and electrical systems. A good specialist heating engineer who is able to prepare reports on heating systems is an added bonus, especially if he also tests and reports upon air conditioning units.

Having organised the office and the equipment and taken instructions from the client, the surveyor must proceed with the survey. He or she will find many varied matters of interest every time he or she leaves the office and considerable challenges. The work will rarely be routine or boring and will often be enjoyable and rewarding, particularly if the advice given can be helpful and the clients who receive it are grateful.

Up to this point, not yet even having undertaken the survey, the reader may have been led to believe that a living which is largely dependent upon building surveys is too hazardous to contemplate. However, if one has a good working knowledge of design and construction, sticks to facts in the report, ignores hearsay and does not venture too many opinions, then building surveys can help to make a rewarding career for the surveyor. Every single building is different, each has its own individual atmosphere and character and there is no better way of appreciating the love and craftsmanship which over the years has gone into so much of our architectural heritage.

Word processors

Word processors are now frequently used for the preparation of reports and most surveyors will be familiar with their operation even if they have not used one themselves. A good word processor, or desk-top computer with word-processing software, can be an invaluable aid to the preparation of reports if correctly used.

In a small office the equipment generally consists of a keyboard, visual display screen and printer with the word processing software and archive software stored on floppy disks or diskettes. Often a number of units are linked together to form an integrated office system, not only for word processing but also for statistical and general computing purposes.

The writer has found that a selection of basic report formats can be stored on floppy disk to cover all the main types and ages of buildings dealt with. These basic reports include all the standard report headings, sub-headings and clauses together with page numbering. When a report is being prepared the appropriate disk is used to present all the basic information applicable to that type of property on the visual display screen, including prompts to remind the surveyor of the particular matters to look out for and cover in his, or her, report.

The information specific to that particular property is then inserted into the basic format on the screen, and specific defects are discussed and then enlarged upon by drawing into the report information stored on disk in the form of 'glossary' text. Thus standard advice on dealing with dry rot, furniture beetle and other defects can be typed into the report by pressing just two keys, attaching first the glossary memory and then inserting text from that memory as required.

The whole process of compiling a report is then simplified and the surveyor may concentrate on the specific issues applicable to the particular building. Finally the report can be read from the screen and adjustments made before it is printed at speeds far greater than manual typing, hopefully free from errors, providing a readable and comprehensive document.

Integrated computers and word processors will allow the storage of file information including details of comparable transactions for statistical and valuation purposes and previous reports highlighting important defects. Information storage by Post Code is a useful way to store and subsequently access property details.

Chapter 2

House surveys

We do not want a structural survey. If you can just have a quick look at the house to make sure that it is structurally sound, that will be fine.
Extract from client's letter of instructions

Inspection procedures

A surveyor should follow a definite procedure when inspecting a property. This procedure will vary from one practitioner to another, but in the writer's view, it is preferable in most cases to begin inside the building at the top and finish outside at the bottom.

For example, the procedure for a conventional two-storey dwelling-house might be:

(1) A preliminary inspection of the whole property to familiarise the surveyor with type and layout.
(2) A detailed inspection of the main roof-space and any subsidiary roof voids which are accessible.
(3) A room-by-room inspection at each floor level starting from the topmost floor.
(4) Inspection of accessible basements, cellars and sub-floor areas.
(5) An examination of the roof structure and coverings from ground level, using binoculars if necessary, and from ladders or through roof voids where accessible.
(6) An examination of the elevations, including structure and finishes.
(7) Inspection of the site boundaries, outbuildings and surroundings.
(8) Examination and testing of the drains.

Detailed procedures

The detailed procedures under each of the preceding sections are as follows:

(1) Preliminaries

The surveyor notes the general character and description of the property, i.e. whether detached, semi-detached, terraced, end-of-terrace, flat, bungalow, two-storey house, three-storey or other. If semi-detached, note if this is a left-hand or right-hand unit, and if terraced or linked to other units the overall length of the terrace should be stated.

The foundations serve a detached structure in isolation and that structure will not normally require provision for movement unless it is very large. Conversely, a long terrace has to be seen as a structural whole and the surveyor must consider whether the entire terrace has been designed to resist the forces to which it will be subject, including possible differential ground and thermal movements. Many examples of terraces of older houses will be encountered, some very long indeed and constructed without any provision for movement. In such circumstances, the inspection should include the elevations of the whole terrace, not just the particular subject property.

Figure 2.1 This massive oak almost touches the garage wing to this surburban semi

The surveyor must note the general lie of the land and the gradient on which the building stands. If the gradient is abnormal there may be a need to consider the possibility of ground movements or landslips and the consequent need for a structural check on the foundation design. Houses built on sloping sites generally suffer more from problems of this type than those built on level sites, and in his subsequent inspection the surveyor will be looking for signs of past structural movements in such circumstances.

Reference may be made to the geological survey map for the area in which the property is located prior to the inspection and this will indicate the likely subsoil conditions. Again, caution is advised if the subsoil indicated on the geological survey map is shrinkable clay, or if worked-out gravel pits or old quarry workings are shown indicating possible unsuitable fill material or made ground. The surveyor will then be aware of the various subsoil conditions in this locality and their characteristics and will approach the survey armed with this information.

An important feature of any survey is to ascertain the age of the structure to be inspected. In most cases the surveyor will know on arrival roughly how old the property is from a first glance. If this is not the case, the surveyor should study the property and those nearby, then ask the vendor, appreciating that he may not be a reliable source of information. The age of the building is most material in deciding the advice to give on a wide range of problems, particularly on pitched and flat roof coverings which have well-defined life spans, as well as the brickwork or stonework, the timbers used, and finishes such as lath-and-plaster ceilings, all of which can vary in life span and quality according to the period during which the building was erected.

(2) The roof voids

The type of roof covering and its condition should be fully described. It may help to draw a sketch plan of the roof and make a note of the size and span of major timber members. At this stage the surveyor will also inspect the ceilings, any ceiling insulation, party walls and gable walls which may be present, and electrical or plumbing services in the roof-space. The detailing around and behind any chimney stacks which may be seen from within the roof should be carefully examined.

It is recommended that a high proportion of the survey time be devoted to the inspection inside the roof since in this way a great deal can be learned about the building. Here one can usually work quietly and undisturbed, except that occasionally a vendor has been known to make use of the surveyor's ladder to carry out his own inspection or even to move items of storage in or out!

Extreme care should be taken in negotiating the ceiling joists if the floor of the roof-space is unboarded. A foot through the vendor's ceiling can cause considerable embarrassment and he will probably assume that the surveyor is a novice. Good illumination is essential, not only for the surveyor to carry out an examination properly but also for his own safety and that of the property under survey.

Surveyors may find to their cost that a leg through an early 19th century lath-and-plaster ceiling is very difficult to extract as, like a shark's jaws, the harder one pulls the tighter the grip of the broken laths becomes.

(3) Room-by-room inspection

It may first be helpful to take the dimensions in each room. A note is then made of the condition of the ceilings, walls and floors. Doors and windows should be tested by opening and closing and any distorted openings and door architraves noted with sloping floors and ceilings or walls which are out of plumb. Fixtures, fittings, fixed appliances, electrical points and other matters of interest should be described.

It is a good idea to roughly draw a sketch plan of each floor level, marking all openings and the locations of power and lighting points, radiators and the like. Such sketch plans, subsequently retained on the file, often prove invaluable in later discussions with clients, particularly when future alterations may be contemplated. If electrical or lighting points are few and far between the client should know about this and he may well not have noticed himself.

Comparison of room dimensions at each level can indicate which are the load-bearing walls and which the partitions which then may be eccentrically loaded on floors. The adequacy of floor construction for spans and loads may then be checked.

(4) Basements, cellars and sub-floor areas

It is always a matter for self-congratulation when the surveyor finds a cellar access under the stairs in an older house, especially if the cellar area disclosed corresponds with most of the ground-floor space. All manner of interesting discoveries about the structure may be made in cellars. Sub-floor ventilation, damp-proof courses, ground-floor timbers, main wall construction, and some of the services may often be capable of close study. As with the main roof void, it is recommended that some time be spent examining conditions in cellar areas and basements. Remember that many of the faults in a building originate, more often than not, at roof or below-ground levels.

Care should be taken to check the condition of the cellar steps since if these are timber and in contact with damp masonry, rot and beetle attack may be expected and defective stair treads may result, particularly if the cellar is disused. Collapse on use has sometimes been known to occur under these conditions.

(5) The roofs

The numbers of chimney flues to each stack should be noted and a check made to trace these through to their fireplaces to confirm if any flues are disused. Redundant flues require ventilated caps or hooded spigots at the top and ventilation at the base in order to keep them dry and prevent condensation forming inside. The general condition of chimney stack flaunching, pointing, brickwork, rendering an oversailing should be noted. The type and condition of the detail at the base of stacks should also be noted, especially behind stacks, and the condition of lead or zinc flashings and soakers or cement fillets described. Tile dentils or cement mortar fillets with no soakers beneath are often unsatisfactory since they tend to pull away from the sides of the stacks in time due to movement in the roof timbers. When this happens, water running down the face of the stacks penetrates directly behind the fillets and causes internal dampness.

The roof covering must be examined, paying particular attention to slipped, cracked or laminated tiles and slates and other signs of deterioration. The condition of hip and ridge tiles is checked, paying particular attention to the bedding mortar which will often be found to be soft and loose. Generally this should not be re-pointed if it is in poor condition but the tiles lifted and re-bedded in new mortar. The condition of valleys is important, particularly secret valleys which may become blocked with leaves and debris.

In older buildings such as Victorian or Edwardian villas and terraced cottages, projecting fire walls may be encountered, rising above the roof line above the party walls. The detailing of these is often poor. The advisability of good flashings and soakers, rather than cement fillets, as for stacks, also applies to this type of parapet construction. A particular check of these should always be made, especially on fire-stop walls, as the coping stones have an alarming tendency to slip off with consequent danger to occupants and perhaps the general public.

A careful inspection of flat roofs, including the feel of such roofs underfoot, is essential. Soft areas of decking, blistering due to trapped vapour, surface cracking to asphalt and built-up felt and other indications of trouble to come should be recorded. The first steps on to a flat roof should be taken carefully since there is a possibility of soft and rotten decking which could be in danger of collapse.

(6) The elevations

A sketch of each elevation may be drawn indicating any specific points of interest. If fractures are present in the walls it is advisable to sketch in all fracture lines roughly, showing starting and finishing points and widths at various points so that a total picture of any structural damage may be given. It is much easier to draw fractures than to describe them in notes, although they will also have to be identified by description in the written report.

Figure 2.2 Thermal movement to the concrete lintel caused slight cracking of the outer skin of this cavity wall

One should stand back and look for any marks or strains on the walls caused by leaking overflows, gutters and downpipes and then mark these points on the sketch. A cross-check made on the internal notes will often show that dampness had been found at the corresponding location inside, especially in older, solid-wall construction.

Many surveyors use photography at this stage and an instant-print type of camera will provide an immediate photographic record of the elevations. These are available for immediate reference on return to the office, whereas conventional photography suffers the drawback of a delay before prints are returned, by which time the report will have been drafted. Unfortunately, fractures do not show up well on photographs when illumination is poor and instant cameras provide only one print.

Inspection of the elevations should include probing of pointing, rendering, brickwork and any cracks or fractures, also examination and probing of exterior woodwork, especially lower window and door frames and exposed fascias. Projecting balcony construction, exposed porches and bays call for particular attention. The elevations should be examined from a distance for signs of past structural distortions or distress and checked, using a spirit level on sills and brickwork courses where appropriate. In older buildings, especially those built with soft mortars, it will be found that considerable structural distortion is able to occur without cracking or fractures becoming apparent.

(7) The site and surroundings

The surveyor studies the general condition of the boundaries and any outbuildings and any comments on these may be framed in accordance with a client's instructions. He could agree with the client at the outset that he will not comment in detail on boundary fences and outbuildings and that the client should satisfy himself regarding these items. Often, from the client's point of view, there seems little to be gained in paying a surveyor for his time needed to inspect and report upon fences, gates, garden sheds, greenhouses and the like. In such cases, and only when the client agrees, a general exclusion clause may be inserted in reports advising clients to inspect these matters themselves in the light of their own particular requirements. The surveyor can then devote time exclusively to the main building and its structural condition. However, particularly in larger, older buildings, permanent outbuildings and extensive masonry boundary walls can form a significant expenditure potential and should then be included in the survey.

At some stage in the report the client should be advised to consult his solicitor regarding the question of liability for boundaries since a boundary fence may or may not be the responsibility of the owner of a particular property, or may be a party fence, depending upon circumstances. This is a matter which should normally be deduced from a study of the title plan and convenants, if available.

As far as the surroundings are concerned, it is suggested that the surveyor checks for any indications of possible neighbouring future development or redevelopment, including road schemes. While the purpose of the survey is not strictly intended to reveal such matters, it is often helpful to comment where appropriate and then to suggest that a client consults his solicitor. Solicitors, for their part, would normally be most grateful for such comments which they can follow up since they will rarely have inspected the properties to be conveyed. If the highway is unmade or is a private road; if the highway is a public road which is narrow or lacking footpaths; these facts could be significant since they may indicate a widening scheme in the future, and in the case of private roads there could be a capital road charge liability in the future if the road were to be made up and adopted.

Should the surveyor notice building works on adjoining land this could be significant. A client may then be advised to investigate the nature of the works proposed, especially having regard to overlooking or to a use which could adversely affect the value of the property he is about to purchase. Such matters would not always come directly to the attention of the client's solicitor in the normal replies to searches and a specific enquiry may be required to unearth the facts if the solicitor is placed on his guard by the surveyor. Similarly, vacant land adjoining the property under review may be destined for development and this possibility would then need further investigation.

(8) Drains

It is debatable whether the testing of drains should be an automatic procedure to be followed in all cases or whether the surveyor should inspect the house and examine the drains with a view to providing a supplementary drains test report later, if required. The latter procedure is recommended.

There are several reasons why initial drain tests may be difficult to undertake. There may be little or no manhole access to the system or the drains could be in such poor condition that any testing under pressure could only cause damage. The drainage system may be part of a long private sewer, the joint liability of all property owners connected thereto. This could require extensive testing some distance from the subject property, perhaps demanding entry within adjoining curtilages. Obviously, if the surveyor were to take a plumber with him in such circumstances, considerable wasted time and an abortive visit could result if tests were not able to be carried out.

I therefore suggest that the surveyor should begin by lifting all accessible manhole covers to check the drain flow and inspect between manholes, using a mirror. He would then make suitable initial comments in the report giving the client an opportunity to commission a supplementary drains test report if needed. This method has a secondary advantages for both surveyor and client since on the basis of the preliminary inspection, the surveyor should be able to assess the extra charges involved in providing such a supplementary report.

Naturally, if pipes are cracked, the falls poor or the drain flow sluggish, such matters can be commented upon initially and suitable advice given.

Manhole covers are frequently damaged, and invariably the older cast-iron covers will be rusted, sometimes solidly rusted to the frame and very difficult to remove without dislodging the frame. If all the covers are rusted solid from years of disuse this does at least indicate that the drains have not required rodding in the recent past.

Clean benching, brickwork and rendering to the inside manhole walls might indicate either that no recent stoppage has occurred or that the chamber has been flushed out after an obstruction. If the drains have blocked in the past and not been hosed out, then water marks and staining to the benching and walls may be found and should place the surveyor on his guard regarding the general drain flow. Blockages can of course occur in a system for quite innocent reasons and may not necessarily indicate a fault in the design or construction of the drainage system.

Occasionally the drains may be completely obstructed. In such circumstances any further inspections would normally be abandoned and the client advised that the system should be cleared, all chambers hosed out and a further visit arranged to prepare a supplementary report. In the first instance it is for the vendor to arrange such matters and the surveyor or his plumber should not be expected to have to deal with drains which are in a blocked or insanitary condition.

It is worth mentioning here that the surveyor and the staff dealing with inspections and tests of drains should take precautions against infection, especially if they have open cuts or abrasions. A periodic tetanus injection is advised in order to maintain immunity. Facilities for washing should be ensured after dealing with drains, handling drain plugs, inflatable bags and other equipment.

100 typical house defects

A list of the one hundred defects which, from the author's experience, are most likely to be encountered is given below. This list is, of course, not intended to be exhaustive but should be of interest. (See also Chapter 7.)

1 Settlement of solid concrete ground floors due to unsuitable or unconsolidated fill material in post-1945 housing. Indicated by gaps beneath skirting boards and dishing to floor slabs.
2 Common furniture beetle (*anobium punctatum*) in roof timbers, floors, staircases and plywood panelling, mostly in pre-1940 housing.
3 Wet-rot (*coniophora puteana*) or dry-rot (*serpula lacrymans*) attack to floors under WC pans due to leaks in trap and flush pipe seals.
4 Wet-rot or dry-rot attack to kitchen or utility room floors due to leaking washing machines or sink traps.
5 Shelling wall plaster in post-1945 houses built hurriedly in wet conditions and plastered on wet blockwork.
6 Failure of lath-and-plaster ceilings in pre-1930 houses, requiring new ceilings.
7 Lamination of machine-made clay roof tiles in pre-1940 houses.
8 Rusting nails to slate roof fixings and failure of slates in pre-1925 houses.
9 Unevenness, dishing and flexing of fully-spanned timber ground floors in modern houses due to excessive warping of joists and chipboard deckings and building down to minimum calculated standards.
10 Leaking built-up mineral felt roofs in post-1945 housing and extensions, often causing wet-rot or dry-rot to timber decking and joists beneath.
11 Inadequately braced roof strutting in conventional purlin timber roofs.
12 Parallelogrammatic movements to roof truss systems which lack diagonal restraint.
13 Rising dampness due to pre-1920 damp-proof courses of felt or slates perishing.
14 Damp penetration due to external bridging of damp-proof courses by soil and paving.
15 Damp penetration due to bridging of damp-proof courses by rubble and debris underneath suspended timber ground floors.
16 Damp penetration due to leaking gutters or blocked rainwater pipes leaking at the collars.
17 Damp penetration due to running overflow pipes.
18 Dampness in solid floors and internal walls due to leaks in embedded central heating pipework.
19 Soft and damp wall plaster due to rain penetration through unrendered solid walls or rendered walls in poor condition.
20 Cracking and movement of block partitions in post-1945 houses where these rest eccentrically on suspended floor construction lacking floor stiffening.
21 Distortion of door architraves in post-1945 houses on upper floors due to the bedding down of eccentrically-loaded partitions.
22 Distortion of door architraves and floors at upper levels in post-1945 houses due to inadequately-supported point loads of copper cylinders, and shrinkage caused by the heat of cylinders.
23 Asbestos cement gutters and rainwater pipes becoming porous with age.
24 Guttering blocked, leaking and broken due to lack of maintenance.
25 Lack of effective ceiling insulation in roof voids.
26 Insufficiently rigid platforms for mounting tanks in roof voids.

27 Galvanised steel hot and cold water tanks rusting.

28 Header tank overflows run into main water tanks.

29 Roof 'spread' due to insufficient collars in conventional roof structures.

30 Foundation settlements due to subsoil erosion from leaking drains.

31 Foundation settlements due to shrinkage of clay soils in the presence of vegetation and certain trees (e.g. poplar).

32 Structural damage due to ground heave on clay soils where buildings have been erected on sites involving the removal of trees.

33 Dampness to solid floor slabs due to punctured polythene membranes in post-1945 houses and extensions.

34 Rusting through of pressed-steel radiators in post-1945 small-bore heating systems.

35 Displacement of lead flashings where there is insufficient depth of chase.

36 Porches pulling away from main buildings due to shallow foundations.

37 Integral garages with inadequate protection against spread of fire into the main building.

38 Loose staircase balustrading, or balustrading and handrails removed making staircases dangerous.

39 Large sheets of ordinary-quality glass in glazed doors causing a hazard to young children.

40 Loose bedding mortar to hip and ridge tiles.

41 Loose and frost-damaged rendering, especially pebbledash and spar-dash.

42 Wet-rot to post-1945 painted softwood window frames and sills, especially to the lowest members; also frequently in porches, conservatories and other extensions.

43 Wet-rot to timber garage doors and door posts.

44 Wet-rot to exposed roof fascia boards.

45 Dry-rot to timber floors and skirting boards adjoining damp walls in pre-1920 houses.

46 Cracked lavatory basins and WC pans.

47 Poor caulking to edges of acrylic baths causing dampness behind bath, especially if shower attachment is used.

48 Poor caulking around shower trays causing damp penetration into walls and floors.

49 Damp penetration around base of, and especially behind, chimney stacks.

50 Leaning chimney stacks requiring rebuilding.

51 Mixed copper and lead plumbing systems causing galvanic corrosion to header tanks.

52 Structural settlement due to building on inadequate ground.

53 Sarking felt overlapped wrongly on roof slopes with lower felt overlapping higher felt.

54 Feed and expansion pipes to header tanks connected wrongly so that header tanks contain hot water and cause condensation in the roof void.

55 Unlagged tanks and plumbing in roof voids.

56 Uncovered tanks in roof voids causing condensation, and dead birds and other contaminants found in tanks.

57 Thatched roofs attacked by squirrels, mice and birds due to lack of protective uPVC or nylon mesh netting.

58 Theft of leadwork from roofs causing leakages.

59 Storm drains blocked due to lack of maintenance and inadequate soakaways.

60 Leaking service pipework when laid at insufficient depth under driveways.
61 Electrical wiring in dangerous condition, possibly including amateur extensions.
62 Insufficient sub-floor ventilation to suspended timber ground floors where sub-floor vents accidentally or deliberately blocked off.
63 Obstructed wall cavities causing bridging and dampness.
64 Leaks from boxed internal soil/vent waste-pipe systems.
65 Condensation damage to decorations and plasterwork in the absence of adequate heating and ventilation.
66 Wet-rot or dry-rot attack to suspended timber ground floor wallplates where in contact with damp masonry.
67 Rot attack to timber lintels and wallplates built into solid exterior walls to elevations facing the prevailing weather or subject to damp penetration from leaking gutters and overflows.
68 Furniture beetle introduced into roof timbers from artefacts stored in lofts.
69 Dampness to inner skin of cavity walls due to incorrect detailing of damp-proof courses, especially when facing the prevailing weather.
70 Water hammer and dripping taps due to perished, worn or incorrect tap washers.
71 Gas leaks, particularly around gas meters and boilers.
72 Drains blocked due to fractured or displaced pipework causing a build-up of solid material.
73 Roof leaks from valleys due to blockage by leaves or other obstructions. Splits to the valley linings often causing rot attack to roof timbers and decking under.
74 Frost attack to common brickwork used in exposed locations on parapets, copings, garden and retaining walls without suitable tanking and weathering and where special-quality brickwork should have been used.
75 Amateur or 'cowboy' loft conversion, with inadequate provisions for escape from fire, fire-resistance and insufficient floor strength.
76 Domestic extensions built without permission or constructed as non-habitable rooms and then converted without permission into habitable rooms having inadequate ceiling heights and/or thermal insulation.
77 Condensation inside sealed double-glazing units due to failure of seal or poor manufacture.
78 Rusting to steel casement windows causing warping, loosening of window stays and catches and cracked glazing.
79 Inadequate bearings for or encasement of steelwork when used for internal alterations or extensions.
80 Warping to interior and exterior joinery, especially thin-section glazed timber doors, hardwood entrance doors and architraves.
81 Chimneys used for gas- or oil-fired boilers without linings causing deterioration to internal wall finishes.
82 Unauthorised changes of use or structural alterations to leasehold property where lessor's consent should have been obtained.
83 Weeping radiator valves causing damp patches to carpets and floors, rust staining to exposed pipes and fittings, and deterioration of floor timbers.
84 Foul drains not airtight due to cracked and ill-fitting manholes with worn covers and frames.
85 Chimney flues used as ducts for electric cable or other services but not sealed.
86 Ventilation blocked or not provided for conventional flue gas appliances.

87 Baths installed without necessary drainage fall to waste outlet causing ponding of water and staining to bath.
88 Poorly fixed or unsafe pull-down ladders to lofts.
89 Incorrect or insufficient weathering and throating to sills and overhangs.
90 Loose parapet copings and coping stones to fire walls in danger of being dislodged.
91 Company stopcocks and gas cocks in front gardens obscured by soil or paving and inaccessible in an emergency. Manhole covers similarly covered with paving or soil.
92 Insufficient ventilation to single-stack soil/vent pipe systems and connecting drainwork.
93 uPVC gutter systems with loose or missing seals, incorrect falls or insufficient brackets causing leaks, sagging and overflows.
94 Over-spanned floor joists in older buildings, often caused when internal walls beneath are removed.
95 Removal of interior walls or chimney work resulting in unsatisfactory slenderness ratio especially when lime mortar is used in walls.
96 Chimney breasts removed at lower levels leaving oversailing breasts at higher levels above with insufficient support.
97 Provision of open-riser staircases, spiral staircases and open stairs without handrails or balustrades in conversion of older buildings and in contravention of building regulations.
98 Use of inadequately fire-resistant materials in ceiling construction.
99 Failure to comply with regulations in respect of flat conversion schemes, especially protected means of escape from upper flats in the event of fire.
100 Lack of suitable intervening ventilated lobby between kitchen and bathroom or WC in converted flats.

Dealing with occupiers

A special feature of the house (or flat) survey which does not arise in other cases is that the surveyor is often entering someone's home when they are still in occupation. Empty houses which are unoccupied and clear of furniture and floor coverings are obviously much easier to inspect but nowadays a majority of house inspections are likely to take place when the house is occupied.

The surveyor in these circumstances has no special rights other than to look and touch.

Great care should be taken in dealing with the occupier's furniture and floor coverings bearing in mind that a liability arises for any damage caused; a surveyor in this context is acting as a servant of the client so that the client would be liable ultimately in law for any damage caused. Few clients would take kindly to receiving a bill for damage caused by their surveyor during the course of his or her inspection.

If suspended timber floors under fitted carpets need to be inspected or if parts of the structure need to be opened-up we would generally suggest that this should only be done with the occupier's written agreement and indemnity. Professional carpet fitters should be used if carpets are of any value. In many cases this will be impracticable initially and a preliminary report will have to be issued advising further exposure work and the issue of a supplementary report at a later stage.

Chapter 3

Foundations

And every one that heareth these sayings of mine, and doeth them not, shall be likened unto a foolish man, which built his house upon sand; and the rain descended, and the floods came, and the winds blew and beat upon that house; and it fell: and great was the fall of it.

St. Matthew VII 24–27

Modern foundation design for simple low-rise domestic structures requires a knowledge of underlying subsoil conditions and loads likely to be imposed upon that subsoil. Foundation design is the province of the specialist structural engineers and while it is considered essential for any surveyor undertaking building surveys to have a full understanding of the principles involved, he should be willing to refer to a structural engineer any doubtful or difficult cases needing further investigation.

I have found it helpful to foster a continuing relationship with a structural engineer's practice. Once such a relationship is established assistance is available, often at short notice, with structural calculations or scrutinising plans and details of an existing building under review.

Design

Modern foundations are normally of one of the following types; shallow strip; deep strip; pad-and-beam; pile-and-beam; or raft. Each type has advantages and disadvantages and will be suitable for some conditions and unsuitable for others. It should be borne in mind that builders generally use the most economical foundation which will satisfy the local authority or current district surveyor, and that fashions in foundation design change. A foundation thought to be suitable for the purpose when laid may later be found less appropriate with the benefit of hind-sight. When the surveyor is aware of details of the foundation design, he should include in his report any comments on that design in the light of current knowledge.

Older buildings often have foundations which fall short of modern standards and are often found to be technically inadequate when checked by calculation. Buildings erected before modern cement concretes became generally available often had very little foundations at all. Common 19th century practice, especially in rural areas, was to provide shallow trenches with a bed of hoggin and to begin the brickwork with two courses of superimposed stepped footings perhaps 500 mm or

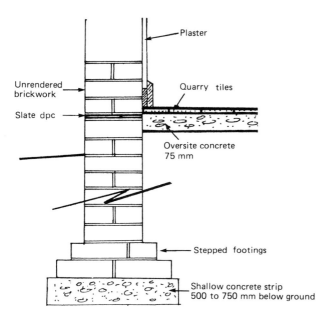

Plaster

Unrendered brickwork

Quarry tiles

Slate dpc

Oversite concrete 75 mm

Stepped footings

Shallow concrete strip 500 to 750 mm below ground

Figure 3.1 Late Victorian foundation, circa 1900

so below ground level, using soft bricks and soft lime mortars. It is surprising how well many buildings erected in this way have stood the test of time but it is sad to reflect that many modern structures, professionally designed and reputably supervised, have suffered foundation failures necessitating expensive repair.

The surveyor will often inspect foundations exposed for this purpose. It is surprising to see how frequently they differ from any original plans and how often the overall depth of the base of foundations is shallower than the design, or prudence, would have required. This will often indicate poor supervision during the course of the works.

A surveyor will not normally be expected to expose the foundations for inspection in a typical building survey. The Practice Note does not require this, it merely states that 'Wall foundations should only be exposed (by the surveyor or an appointed contractor) in cases where structural defects indicate problems below ground, after prior approval'.

The results of a foundation inspection may be relayed to the client in the form of a supplementary report, if necessary, and if the client should commission it. In some cases the surveyor may be able to arrange for inspection pits to be dug and the results reported. In other cases, especially where abnormal ground conditions may be present, the surveyor should undertake foundation investigations in conjunction with a structural or geotechnical engineer, particularly if the results of trial bores and recommendations need to be conveyed to the client.

The surveyor must not be afraid of stating his views on a foundation when this is inspected and he is sure of his ground (in both senses), neither should he be reluctant to refer to a specialist where this seems necessary in the client's interests.

The Building Regulations lay down the following standard for foundations:

Foundations must (a) safely transfer all dead, imposed and wind loads to the ground without settlement or other movement which would impair the stability of, or cause damage to, the building or any adjoining building or works; (b) be taken down below frost damage or soil movement levels; (c) be resistant to attacks by sulphates or other deleterious matters in the subsoil.

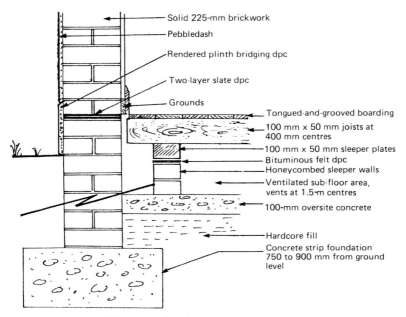

Solid 225-mm brickwork
Pebbledash
Rendered plinth bridging dpc
Two-layer slate dpc
Grounds
Tongued-and-grooved boarding
100 mm x 50 mm joists at 400 mm centres
100 mm x 50 mm sleeper plates
Bituminous felt dpc
Honeycombed sleeper walls
Ventilated sub-floor area, vents at 1.5-m centres
100-mm oversite concrete
Hardcore fill
Concrete strip foundation 750 to 900 mm from ground level

Figure 3.2 Typical 1930 domestic foundation

It is suggested that the foundations of any structure inspected and reported upon by the surveyor should be judged by this standard and if this is not met or there is doubt, this should be discussed in the report with the implications set out clearly.

In England and Wales a depth of 450 mm is sufficient to place foundations below the level of frost damage in the most extreme conditions for which provision might reasonably be expected to be made. In practice it is felt that a greater depth than this is generally advisable because of the likelihood that ground conditions at the surface on a site may vary, and areas of topsoil or old excavations could be found which would make the bases of shallow foundation trenches subject to variations in bearing capacity. Accordingly I would generally regard 900 mm as being more suitable. Any foundation shallower than this must be regarded as being less than ideal unless the bearing is a hard rock formation, or similar base, of undisputed strength and regularity.

Subsoils

Care is required on chalk soils to ensure that the foundations are at least below the minimum for frost damage since chalks can be subject to considerable frost heave close to the surface.

Despite Biblical claims to the contrary, sandy soils make good foundation bases in many cases, although high-frequency vibrations in the range of 500 to 2500 impulses per minute can cause serious settlement problems, a point unlikely to arise with residential structures unless near to machinery. This factor should be borne in mind when dealing with industrial structures, particularly in the case of changes of use involving machinery.

On shrinkable clay soils a foundation depth of 1200 mm is recommended which would be below the level of seasonal change on level sites in the absence of vegetation. On sloping sites rather greater depths are required, especially on south-facing slopes where the ground will drain and dry out readily during periods of dry weather.

The presence of trees or bushes on shrinkable clay sites causes a particular problem for surveyors advising a prospective purchaser. The standard against which the property should be judged should be that set out in the appropriate British Standard. (BS 5837:1980 'Code of Practice for Trees in Relation to Construction'). This British Standard states, 'Where the presence of shrinkable clay has been established and where roots have been found or are anticipated in the future, it is advisable to take precautionary measures in the design of the building's foundation (in accordance with CP 2004:1972 and CP 101:1972). Alternatively, consideration should be given to designing the superstructure to accommodate any foundation movement that might be induced by clay shrinkage. Where a subsoil investigation has not been carried out, an approximate rule-of-thumb guide is that on shrinkable clay, if risk of damage is to be minimised, special precautionary measures with foundations or superstructure should be considered where the height or anticipated height of the tree exceeds its distance from the building'.

One should pause for a moment to consider the full implications of BS 5837. In most cases the property being inspected, whether residential, commercial or industrial in character, is unlikely to have special foundations of the type anticipated, unless it is quite modern. Generally a short-bored pile-and-beam foundation, or pad-and-beam, or occasionally a raft design, would be used to accommodate the seasonal ground movements induced on clay sites by tree roots. (A sketch section through a pile-and-beam foundation is shown in *Fig. 3.3*). It follows that for most older structures on clay sites a traditional shallow foundation design may be assumed, and that if trees are present whose height at maturity will exceed their distance from the walls of the structure, then the structure is at risk from shrinkage of the clay subsoil and damaging foundation movements would be possible during a dry summer.

Accordingly the surveyor should make a note of the locations of all trees and bushes near the building being inspected and record their distances from the main walls, and that he identify the species of tree involved, confirming whether or not the subsoil may be shrinkable. It should also be confirmed whether specific trees or bushes may be within the curtilage of the property under review, or in adjoining curtilages.

The surveyor should then comment and warn the client as necessary in the light of current knowledge. Three separate situations may arise. A building with shallow foundations on shrinkable clay may have trees within the curtilage closer to the structure than their mature height. In similar circumstances there may be such trees but they may be located in an adjoining curtilage under other ownership. Possibly there may be trees within the curtilage of the subject property which, while well back from the property concerned, could be a threat to an adjoining structure and thus give rise to a legal liability in the future.

If trees are closer to the building than the recommended distances, and within the curtilage, some action in the nature of removal, crowning, pollarding or simply

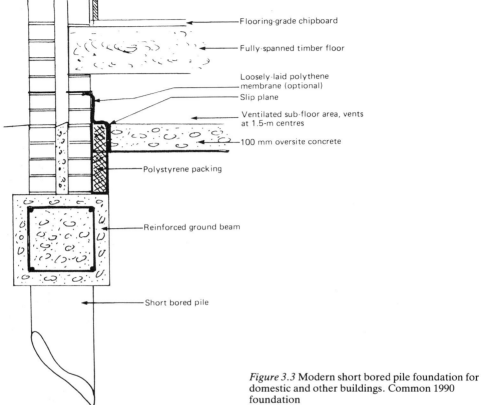

Flooring-grade chipboard

Fully-spanned timber floor

Loosely-laid polythene membrane (optional)

Slip plane

Ventilated sub-floor area, vents at 1.5-m centres

100 mm oversite concrete

Polystyrene packing

Reinforced ground beam

Short bored pile

Figure 3.3 Modern short bored pile foundation for domestic and other buildings. Common 1990 foundation

subjecting to a period of observation and future pruning, may be advisable depending upon the circumstances.

If trees in other ownership are closer than would be recommended then the surveyor could advise the client to write to the owners of these trees pointing out that their root systems could be hazardous to the property he is acquiring and suggesting that further advice be taken in the light of the fact that by virtue of the laws of nuisance the owner of a tree is liable in law for damage caused by its roots extracting moisture from the subsoil under adjoining structures. (See *Bunclark* v *Hertfordshire County Council* (1977) 243 EG 455 reported in *Estates Gazette*, 30 July, 1977).

Conversely, in a situation where trees within the curtilage of the property being acquired could pose a threat to adjoining structures, then the client should be warned of his own potential liability under the law of nuisance in case this should become significant in future years.

Certain species of tree are really unsuitable in a garden setting on shrinkable clay soil, particularly poplars, willows and elm, and if these are found they should be regarded with a critical eye. Poplar roots can extend for distances of up to twice the height of the tree in search of moisture. I have also found ash trees to have searching root systems and a particular liking for entering cracked drains and disturbing paving, quite apart from their possible effects on shallow foundations.

It is recommended that surveyors practising in localities where shrinkage clays are found should familiarise themselves with the various aspects of this particular problem and that they should be able to identify any species of tree encountered during their work where this seems necessary. The reader may find *Trees in Britain, Europe and North America* by Roger Phillips (Pan Books) to be a most useful guide to tree identification throughout the seasons.

One factor which can commonly cause problems with older foundations is subsoil erosion caused by cracked or leaking drains, or leaking water pipes. Unfortunately, storm drains and gullies to older property were rarely constructed well and soakaways will often have become blocked after the passage of time so that a point-discharge of water into the subsoil immediately adjoining the main walls will often occur. This can be damaging, especially to foundations on light sandy or silty soils which erode easily. Clients should be warned of the importance of maintaining both foulwater and stormwater gullies and drains in good order and to avoid any point-discharges of water into the subsoil adjoining the main walls.

Good building practice is to maintain foulwater drains in a watertight condition in normal use and to lay storm drains watertight for a distance of 3 m from any structure. After this distance, storm drains may be laid as open-jointed land drains if required, and may run to soakaways if alternative stormwater sewers are unavailable and the soil is suitably porous.

Clients will rarely give a thought to what happens to the storm and foulwater discharges from their property and their significance unless these points are described and their importance to the structure is discussed. In practice, lack of access to storm drains will often preclude any description of the actual system by the surveyor.

Figure 3.4 Classic settlement – the footing has dropped, causing cracks which widen with height

Assessment of damage

Foundation defects requiring a specific remedy are those which cause serious distortion or separation in the structure on a scale which would be unacceptable to a reasonable occupier. Specific remedies would deal with foundations which are (a) not deep enough, or (b) not wide enough, or (c) not firm enough, for the particular subsoil conditions involved. The traditional remedy for such defects would be to underpin foundations in such a way as to make the overall structural support to the footings either deeper, wider or firmer as needed.

Figure 3.5 The London plane trees in the pavement have caused shrinkage of the clay under the shallow bay foundations

First one must consider those cases where such work would be unnecessary or premature. A number of specialist underpinning contractors exist who undertake traditional or system methods using interrupted foundations of various types. Such contractors will be commercially-orientated and will wish to emphasise the advantages of underpinning using their own particular system. A surveyor will need to consider whether or not it is necessarily in his client's best interests to refer a problem to such a contractor rather than to a professionally-independent structural engineer.

The first matter to be considered would be available evidence of actual past and present damage to the structure under review, having regard to age and all circumstances. Initially, any above-ground damage would need to be assessed, for example, cracks in walls (inside and outside), evidence of distortion to window and door openings, sloping floors and irregular ceilings. The BRE current paper 'Foundations for Low-rise Buildings', CP 61/78, and 'Assessment of Damage in

Low-rise Buildings' in *BRE Digest* 251, make certain recommendations for the assessment of crack damage as follows:

Category 1   *Degree:* Very slight
Fine cracks up to 1 mm wide which can be treated during normal decoration.

Category 2   *Degree:* Slight
Cracks up to 5 mm wide, some external repointing may be required to ensure weather-tightness. Doors and windows may stick slightly.

Category 3   *Degree:* Moderate
Cracks between 5 and 15 mm wide, or a number of cracks up to 3 mm wide. Cracks require opening up and stitch-bonding by a mason, with repointing to exterior brickwork. Doors and windows sticking. Service pipes may fracture. Weather-tightness often impaired.

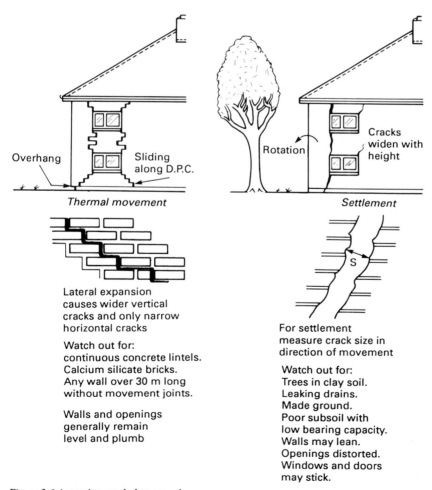

Figure 3.6 Assessing crack damage – 1

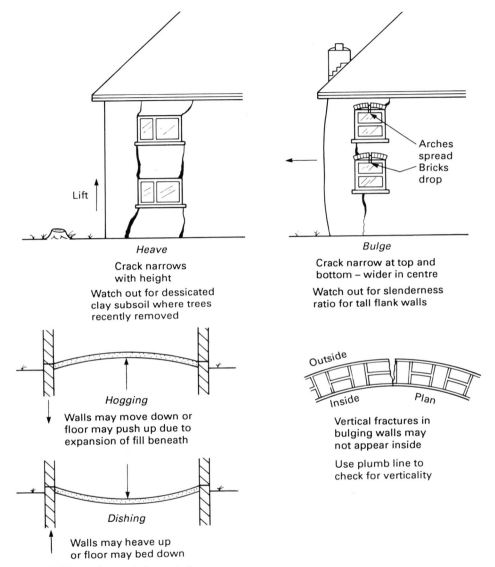

Figure 3.7 Assessing crack damage – 2

Category 4   *Degree:* Severe
  Cracks between 15 and 25 mm wide, but also depending upon numbers of cracks. Extensive repairs needed to sections of walls. Window and door frames distorted, floors sloping, walls leaning or bulging. Some loss of bearing to beams. Service pipes disrupted.
Category 5   *Degree:* Very Severe
  Usually 25 mm or more but depending on number of cracks. Requires major repair involving partial or complete rebuilding. Beams loose bearings. Walls lean badly and need shoring. Window glass cracks. Danger of instability.

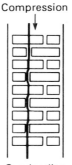

Compression

Overloading

Vertical cracks to
piers indicate
compression failure
with wall eventually
bursting laterally

Watch out for:
Slender brick piers,
especially at
corners and bays
taking high loads.
Weak stonework.

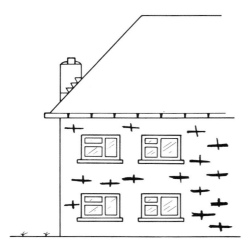

Rusting wall ties

Pattern of horizontal
cracking to mortar
joints in cavity wall.
There may be bulging
and separation of two
skins of brickwork.

Watch out for:
Any pre-war cavity walls.
Black ash mortar.
High exposure to driving rain.
1950sFinlock gutters with
leaking joints.

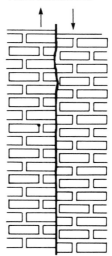

Shear failure

Often to bonding
between new and
pre-existing construction

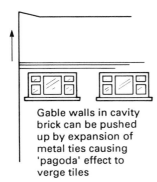

Gable walls in cavity
brick can be pushed
up by expansion of
metal ties causing
'pagoda' effect to
verge tiles

Figure 3.8 Assessing crack damage – 3

Damage of Categories 1 and 2 proportions would not generally of itself justify any specific remedy, and a cosmetic repair should be sufficient. Categories 1 and 2 damage in circumstances where other adverse features of the structure or site exist may, however, be a foretaste of more serious problems to come. Most major structural fractures start with a hairline crack.

Category 3 damage could require a specific remedy depending upon the circumstances. Certainly such damage would require periodic monitoring in order to confirm whether or not the structural movement is continuous and worsening.

Categories 4 and 5 damage will probably require a specific remedy in addition to above-ground repairs. There is a likely need for some type of underpinning in these cases.

It should be mentioned at this stage that some settlement in certain structures is quite normal, being due either to the initial consolidation of the building after construction or to long-term consolidation over the life-cycle. A BRE current paper, 'Settlement of a Brick Dwellinghouse on Heavy Clay 1951–1973', CP 37/74, on this topic covered a 20-year history of settlement in the narrow strip foundations of a typical house in Hemel Hempstead, Hertfordshire. This settlement was found to continue fairly steadily for the whole period (without causing any cracking to the walls) at rates of between 0.12 and 0.39 mm/year, with angular distortion of the foundations to 0.1%. More-brittle wall surfaces, especially surfaces with a smooth stucco, would have cracked as a result of such angular distortion causing Categories 1 or 2 damage, but of itself this would not have been significant.

The question of which category in which to place a specific observed area of damage is very much a matter of judgment. A large number of narrow cracks may,

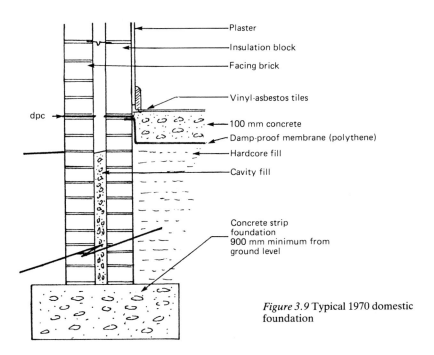

Figure 3.9 Typical 1970 domestic foundation

Figure 3.10 Sometimes the removal of a tree can cause upward heave to clay soils

Figure 3.11 With some cracks it is not immediately obvious what the direction of movement is

in some cases, be more significant than a few larger ones. The recommendations for categorising damage can be no more than a guide, and each property and area of damage has to be individually assessed on its merits. Having decided initially on the extent of the damage, advice must then be given.

There are three possible situations with which the surveyor may have to deal. First, he may be acting for an intending purchaser and in this circumstance the client will have the choice of either buying the property or not buying it. If he decides to buy he will probably have to do so within a reasonably short space of time, therefore there will be no opportunity for an extended period of observation. If there is any doubt about the significance of structural damage, it is clearly better *not* to buy the property and to look elsewhere. If then having had all the possibilities explained to him the client still wishes to buy, it is suggested that some adjustment of the purchase price might be recommended having regard to (a) the maximum likely cost of remedying the most severe defect likely to exist, and (b) the probability that such a remedy may be required having regard to the results of whatever investigations the client is prepared to undertake prior to buying.

The second situation occurs when the surveyor is consulted by an owner who is worried by signs of structural distress in his property. Advice given might then be (a) to do nothing, or (b) to subject the structure above ground to an extended period of observation using tell-tales and other indicators, or (c), in Categories 4 and 5 cases, to undertake temporary shoring and carry out trial excavations to determine the facts with a view to a subsequent specific remedy such as underpinning.

The third situation which can arise is where a building is inspected for a prospective mortgagee and signs of structural distress are found. In this situation the amount of the mortgage advance proposed, in relation to the current market value, should be taken into account and a view expressed on the client's behalf of structural repair costs which could arise in relation to the probability of such repairs being needed and to the percentage advance involved. In many cases of lower percentage advances it would be reasonable to make such an advance on the security of a building suffering Categories 1, 2 or 3 damage. Categories 4 or 5 damage would require further investigation unless the mortgage advance was very low indeed. In addition to advising mortgagees in such circumstances the surveyor would be bound to bring such a situation to the attention of the borrower if he is to have a copy of the report, and the borrower should be advised to take further advice before proceeding, irrespective of the outcome of the mortgage application.

A further point worthy of note in connection with structural damage due to subsidence, landslip or ground heave (normal insured risks in many building insurance policies) is that such a cover applies only to damage arising after the policy came into force. It follows that an existing building owner may be covered for such risks in respect of damage apparent in his building but that if he sells, the purchaser may *not* be covered if signs of damage were apparent before the sale. Once a purchaser has commissioned a surveyor's report which has described structural damage he may be placed in a different position to recover subsequent costs under his insurance policy. In such a situation it may be that the vendor could make a claim and agree the extent of remedial works required prior to any sale. A purchaser should not assume that, because he is insuring against such matters, minor cracks reported by his surveyor need not concern him since the very fact that the surveyor has reported the damage may have undermined the basis of any later claim.

Remedies

Damage caused by the action of tree roots and proved to be due solely to the presence of trees and no other cause should not normally require underground repair. It should be sufficient to recommend that the offending trees be removed and that the site be given a number of wet winters to stabilise before the damage is made good by stitch-bonding cracked brickwork, repairing pointing and making good rendered surfaces in the normal way. If the offending trees cannot be removed, because they are either protected by tree preservation orders or in other ownership, then deeper foundation bases may be required to accommodate present

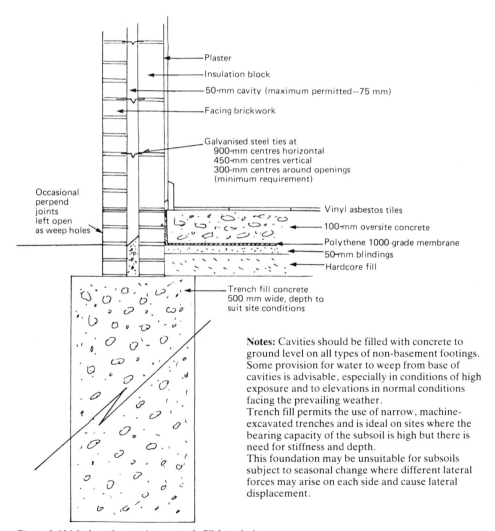

Plaster

Insulation block

50-mm cavity (maximum permitted—75 mm)

Facing brickwork

Galvanised steel ties at
900-mm centres horizontal
450-mm centres vertical
300-mm centres around openings
(minimum requirement)

Occasional
perpend
joints
left open
as weep holes

Vinyl asbestos tiles

100-mm oversite concrete

Polythene 1000-grade membrane

50-mm blindings

Hardcore fill

Trench fill concrete
500 mm wide, depth to
suit site conditions

Notes: Cavities should be filled with concrete to ground level on all types of non-basement footings. Some provision for water to weep from base of cavities is advisable, especially in conditions of high exposure and to elevations in normal conditions facing the prevailing weather.
Trench fill permits the use of narrow, machine-excavated trenches and is ideal on sites where the bearing capacity of the subsoil is high but there is need for stiffness and depth.
This foundation may be unsuitable for subsoils subject to seasonal change where different lateral forces may arise on each side and cause lateral displacement.

Figure 3.12 Modern deep strip or trench-fill foundation

and future root activity. An interrupted foundation should be used to ensure that expansion and shrinkage rates in the subsoil are the same on all sides of the foundation at different times of the year. If trees are a contributory cause of damage but the foundations are also inadequate for other reasons then an underground repair may be necessary in addition to the removal, pollarding or crowning of offending trees.

Traditional underpinning, which is an expensive and labour-intensive process, comprises the provision of an additional foundation below the existing foundation. This may be achieved by alternately excavating bases, usually $1\,m^3$ in extent, around the perimeter of the structure and under the internal load-bearing walls. Each base is cast incorporating steel reinforcement married into adjoining alternate bases so that a complete continuous foundation is provided under the existing foundation to whatever dimensions are specified. When the bases have hardened, the space between new and old foundations is filled by dry-packing with a hard cement mortar. During the course of such underpinning, allowances have to be made for diverting and repairing any services encountered. After the work is finished, a suitable period is allowed for consolidation of the structure on the new foundation, and then any cosmetic repairs, masonry repairs and levelling up are completed.

As a general rule such underpinning will be applied equally to all bearing walls, otherwise a danger of structural movement between those parts of the building which have been underpinned, and those parts which have not, arises. This may not apply on a sloping site where the underpinning depth may step up with the site slope and remain in the same stratum of subsoil.

Special difficulty arises with semi-detached or terraced structures in multi-ownership since if the general rule mentioned is to apply, the whole structure or terrace would need underpinning. In such situations a degree of engineering judgment is required and it may be necessary to undertake partial underpinning only. If so, it is recommended that the depth of the underpinning bases be graduated down from the point where non-underpinned walls end to avoid a sudden marked change in overall foundation depth at any specific point. If party walls have to be underpinned, hopefully this will be by agreement since although a party wall can be physically underpinned from one side only, the bases themselves will be trespassing beneath and adjoining ownership and permission is then needed. There is also a risk that partial underpinning of, say, a semi-detached house and the party wall will result in any future stresses being transferred to the adjoining house. This could cause movement relative to the party wall with consequential damage. Before advising on such works legal advice should be taken as to the implications and means by which the adjoining owner might be warned about the proposed works. A party wall may be underpinned by Notice under the London Building Acts (Amendment) Act 1929, part of the special code for party structures which applies in London now being extended to England and Wales.

As an alternative to the traditional underpinning described, a number of specialist contractors now employ special systems to provide greater depths using a system of pads and beams. Pads are cast at corners and within the structure at the main point-loads to considerable depths if necessary. The pads are then linked by narrow reinforced ground beams cast directly under the original foundations. Such systems reduce the overall amount of excavation necessary and provide interrupted foundations which could be recommended on certain types of shrinkable clay, especially if live tree roots are present in close proximity.

If a surveyor is asked to deal with underpinning he or she should be satisfied that it is necessary in the circumstances that the shoring and other temporary support works will be adequate and that the procedure for casting a new foundation will not subject the structure or any adjoining structure, to additional distress.

Generally, with traditional underpinning, bases should be cast leaving at least two equivalent areas of support between so that bases 1, 4 and 7 would be cast, allowed to harden, dry-packed completely and the packing set. Then bases 2, 5 and 8 might be excavated similarly, followed by bases 3, 6 and 9. In soft ground even greater spacing of excavations may be required to avoid further structural damage. In all cases excavations should be open for the least possible time since some subsoils deteriorate when exposed. All excavations should be adequately planked and strutted at all stages of the work to avoid lateral displacement of the adjacent subsoil and the bottom of any underpinning excavation should be carefully checked to ensure that it is sound, level and dry before concrete placement begins.

If it is suspected that the structure under review has been underpinned in the past, the reasons for this should be ascertained and details obtained if possible. Nowadays, underpinning works require local authority consent under the Building Regulations and most recent works should appear in their records. Before 1976, such work was regarded as outside the scope of the Building Regulations or bye laws and normally went unreported to the local authority.

Drains and foundations

During the course of the survey it may be found that drains pass beneath foundations and floors or a shallow drain may pass through the walls below ground level and beneath the floors.

Certain precautions have to be taken when building over drains. Occasionally, a structure may have been built over drains originally and this will be found in many terraced Victorian and Edwardian houses with private drains running under the buildings from rear to front. Often, extensions to an existing building will have been erected over the drains at a later stage and this will commonly be found in small, single-storey domestic extensions.

Hopefully a manhole may be found on each side of the structure where the drain passes beneath and it should be possible to test the drain for leakage and to inspect the pipework for distortion using a mirror. Such drainwork should be laid with a compressible layer above and immediately over the pipework and with suitable lintels used to carry strip foundations or footings over the pipes without imposing a point-load. Modern uPVC underground drain systems are sufficiently flexible to cope with slight distortion without damage but older virtrified earthenware pipework is brittle and could fracture under compression caused by foundations.

One important point which should be considered in drainwork and shallow foundations is the relative depths of the original excavations where these would have been made side by side. It is not uncommon to find that the invert level of drainwork is deeper than the base of strip foundations. If the two excavations proceeded together at the time the building was erected, or if the drains were laid later and if the excavations have been in close proximity, then there would have been a danger of lateral displacement of the subsoil into the lower excavation following the work and during the period of back-fill consolidation. This could have an adverse effect on the subsoil bearing the weight of the adjoining foundation, and the effects may last for some years causing settlements in the structure adjoining the drains until finally the subsoil will have fully consolidated to a point where further movement will not be significant.

It is therefore recommended that invert depths of all drainwork near walls of buildings be checked where possible and especially if that drainwork runs alongside walls. Deep invert levels adjoining structures built with shallow foundations may indicate that subsoil consolidation around the back-filling could be taking place with adverse effects on that structure.

Mining and filled ground

Some surveyors practise in mining areas which are either active or worked out. In localities subject to mining subsidence it may be necessary to employ specialist advice before reporting on buildings, and it is recommended that the surveyor's library should include all available information in the form of plans and maps showing areas subject to mining where these are known. Coal-mining areas are mainly affected by this problem but there are many other types of mining and some areas have a mining history which goes back to the Middle Ages and beyond. In such situations the local knowledge of the surveyor may prove invaluable.

In areas having a chalk subsoil old chalk-mine workings may exist and the chalk itself may be subject to the sudden unexpected appearance of deep swallow holes either connected with workings or due to subterranean collapse caused by groundwater erosion. There is no effective way in which a surveyor carrying out a normal survey can forecast such events. The only protection available for the property owner against such eventualities would be to carry the normal subsidence and landslip insurance cover.

In some localities chalk pits, gravel pits, sand pits and worked-out quarries may be encountered where materials have been removed and the workings filled using either hardcore or compressible fill material. Buildings erected on bad ground where unsuitable fill material has been used, or where suitable material has not fully consolidated, are often found to be subject to movement. This type of structural movement can be highly damaging, occasionally requiring the demolition of the building and often necessitating a special underpinning scheme making piles (or deep pads) to take foundation loads below the level of the fill. Because of the pressures to build on existing urban sites or reclaimed industrial land, the numbers of serious defects attributed to building on made ground appear to have increased in recent years.

An Ordnance Survey map of the area sometimes indicates old mines and workings, especially on $1/1250$ and $1/2500$ scales. This can be a useful source of information, also the geological survey map may show where such workings have taken place. Old maps of the area sometimes show detail which no longer appears on modern maps and the writers have found the old Ordnance Survey county series of large scale maps to be useful, sometimes in showing conditions on a particular site before it was developed. Many of the older county series maps date from the last century. The old $1/2500$ scale maps are often very accurate and informative.

The vast majority of foundations will perform satisfactorily throughout the life of a building notwithstanding the apparently inexhaustible range of foundation problems which could, in theory, arise.

A great deal of superficial damage to buildings, especially cracking, is minor and due to normal long-term consolidation of the structure or minor differential settlement. This must be regarded as normal, especially in older buildings. Some damage may be wrongly diagnosed as due to foundation movements when other and more prosaic causes may be responsible, especially in the case of widely-based structures where thermal movements may be expected and in recently-completed buildings where normal shrinkages, bedding down of partition walls and the like, will produce considerable cracking which may be quite unrelated to conditions below ground.

Serious damage is that which may be diagnosed as progressive and which will require a structural repair in order to arrest further movement. Most cracking is minor and not progressive. In the case of progressive movement there will normally be indications of the cause and this will often be obvious to one who has a knowledge of local soil conditions and the presence or otherwise of mining or made ground in the locality.

In certain circumstances compensation is payable for damage caused by mining subsidence if this can be shown to arise from the activities of the National Coal Board. Subsidence is also normally an insured peril in a standard comprehensive household insurance policy, but not in all cases and frequently not at all in policies on commercial buildings. Insurance cover will not normally include defects which existed or originated before the policy was taken out.

Landslips and coastal erosion

The surveyor with local knowledge is often at an advantage in those localities where there are fault lines, slip planes or coastal erosion. Some districts are notorious for the damage caused to buildings by continuing shifts in the landscape; for example, at Ventnor on the Isle of Wight a local fault line (or 'vent') is responsible for a long history of structural damage to buildings constructed on the boundary between two geological masses. Slip planes are found between layers of clay or other material on hilly or undulating sites and one mass of clay will be sliding over a mass below causing highly damaging shifts to foundations (see Morgan v. Perry in Chapter 19).

If a building is located close to a cliff or headland then clearly it is necessary to consider the possibility of coastal erosion and if there is any doubt regarding the condition of any coastal defences or the stability of the local geology then the client would be well advised to steer clear of such a property.

Chapter 4

Walls

Now the wolf, having blown down the house made of straw and....having blown down the
house made of sticks...., came to the house made of bricks with the third little pig inside.
'Little pig, little pig', he said, 'May I come in?' 'By the hairs on my chinny chin chin', said
the third little pig, 'You may not come in'. 'Then', said the wolf, 'I shall huff and I shall
puff and I shall blow your house down....So he huffed and he puffed....but he could not
blow down the house....because it was built of bricks.
The Three Little Pigs, Anon.

During a survey the thicknesses, lengths and locations of the walls around and
within the building should be noted, including any abutting or connecting walls
which are part of adjoining premises or boundaries. Those walls which are
load-bearing walls and those which are non-load-bearing walls can then be
distinguished one from another.

In simple two-storey dwelling house construction, as in the ubiquitous
between-the-Wars semi-detached, there is usually a central cross wall which
supports not only its own weight but also the loading from floor and ceiling joists
which span front-to-rear in the structure, and perhaps the strutting to roof purlins.
Openings may have been formed in such a wall and a check on the adequacy or
otherwise of remaining piers and bearings is recommended if two living rooms have
now been formed into a 'through-lounge'. It must be borne in mind that 150 mm of
bearing is desirable for most short steel sections and that in a 225-mm solid brick
party wall only 112 mm of bearing is technically available each side of the centre
line. Often there arises a temptation to dispense with piers with the result that the
bearing for rolled-steel joists in through-lounge conversions is often inadequate.

With a dimensioned sketch plan of the room layout on each floor one can easily
see which walls rise continuously within the building and which rest eccentrically on
the floors. In the inter-War structure already mentioned it is likely that at least one
main cross wall is provided and that some other walls are eccentrically-loaded on
the floors using a double-joist for support if the joist runs are parallel to the wall, or
a sole plate fixed over the boarding if the wall is at right-angles to the joist runs.

In post-War construction a greater use was made of fully-spanned roof trusses
and trussed rafters so that often all the partitions to first floor rooms rest on floors
and do not correspond to the location of ground-floor walls. Such partitions may be
timber-framed, consisting of studs and noggins clad in plasterboard and therefore
imposing fairly minimal loads, or they may be of blockwork resting on sole plates

44

on a boarded or chipboard floor, imposing some point loads on the floors. Sometimes considerable deflection will occur when partition walls of this type rest on the floor construction. A particular watch needs to be kept for such signs, especially in modern speculative construction. If the partitions do not rest centrally over the double joists beneath, or if joists have been omitted, or if a series of point loads coincide, then deflection, twisting and leaning of partition walls will occur and the solutions, though simple in theory, in practice often involve considerable disruption to the occupants.

The effects of joinery shrinkage in floors can be seen in deflection of partition walls resting on those floors, and in any event timber floors will tend to deflect downwards progressively with age. If, when the surveyor examines the architraves to doorways on the landing of a house incorporating this type of partitioning, he notices some slope to the tops of those architraves, he will know that movement has occurred. Generally in these cases downward slopes away from the main cross wall towards the partitions will be apparent.

Within normal tolerances such sloping door architraves and minor plaster cracking over doorways may be regarded as normal and an opportunity should be taken to give the client some reassurance that the cause is downward deflection of the supporting floor construction and not something more fundamental lower down in the building.

Point loads of copper cylinders and airing cupboards often contribute to this type of movement in partition walls and door architraves, both from the heat generated causing shrinkage, and the weights imposed. Good building practice should provide for two 100 mm × 50 mm timber bearers under copper cylinders to give at least 50 mm of clearance from the floor with the bearers spanning at least two, and preferably three, floor joists. The National House-Building Council requires such a 50-mm clearance in all new construction in order to meet their standards. The surveyor should check the hot water cylinder bearings. If the cylinder rests near or over a supporting wall beneath then the bearing may not be significant, but if it lies mid-way over a ceiling span to the floor beneath this may be important. Similar considerations also relate to the platforms on which cold water cisterns and header tanks are mounted since they may affect the ceiling construction beneath.

In dealing with older houses of up to Victorian or Edwardian times the importance of taking full notes showing the arrangment of the walls and their construction is even greater since there are many examples of such buildings where internal walls to upper storeys are all timber-framed stud partitions having very little resistance to tension at the points of contact with the main outer walls. The surveyor should then consider whether the outer walls are too slender and lacking in lateral restraint or buttressing to an extent where a danger of structural movement arises. A careful examination should then be made for evidence of bulging and leaning, especially at tall gable ends and other similarly sensitive points.

Frequently, tie bars with S-plates or X-plates have been employed to restrain outer walls in older buildings either as a precaution or where separation has begun. If possible the condition of the tie bars, which often run under the floors or over the ceiling joists in the roof space, should be checked. If these are old the mild steel may have rusted and their effectiveness could be lost.

Tie bars and similar means of holding together old buildings which may lack any internal restraint and may be built in soft lime mortar must be regarded as unsatisfactory by comparison with a better initial standard of overall construction. Nevertheless, once the surveyor has warned his client of the limitations of tie bars and the inherent drawbacks of a building which needed them in the first place, they have often been found to serve their purpose well. Their use is still advised on occasions where this seems sensible and having regard to the depth of the client's pocket at the time.

Figure 4.1 It is imprudential to allow a gutter to discharge down the face of a wall

A cautionary sign arises when it is found that chimney breasts have been removed from existing structures. In old buildings the chimney breasts often act as piers, providing valuable stiffness, and their removal is not recommended unless a careful check on the surroundings indicates quite clearly that the remaining construction will be unaffected, having regard to the slenderness ratios of remaining walls and the nature of the mortars used and therefore their resistance to tensile forces.

Slenderness ratio

Slenderness ratio is a useful concept to use in assessing the structural stability of buildings under review. It is recommended that the standard to apply to old buildings is Regulation D15 and Schedule 7 of the old Building Regulations for structural works of bricks, blocks or plain concrete, and using Schedule 7 it is not necessary to calculate loads or wall thicknesses required, the rules of the Schedule being sufficient as a check on the stability of the construction.

Schedule 7 may be applied to walls which form part of any storey of a building other than a basement, basement storey walls needing a structural check by calculation in accordance with BSI CP 111:1970 should this be considered necessary. It is assumed for this purpose that point-loads are properly distributed and that walls are properly bonded and solidly put together. There is an overall height limit of 12 m for the use of these rules (above which CP 111 should be used) and also a limit of 3 kN/m² as an imposed floor load. This is well above the imposed loads likely in residential construction but which could well be exceeded in other types of building, in which case CP 111 and special calculation would be required.

The Table to Rule 7 sets out certain minimum thicknesses for external walls and separating walls, in each case assuming that walls will have at each end either a pier, buttress, buttressing wall or chimney unless the wall concerned forms part of a bay window and is less than 2.5 m in both height and length. This Table is set out below:

Height of Wall	Length of Wall	Minimum Thickness of Wall
Not exceeding 3.6 m	Any length	200 mm throughout its height
Exceeding 3.6 m and not exceeding 9 m	Not exceeding 9 m	200 mm throughout its height
Exceeding 3.6 m and not exceeding 9 m	Exceeding 9 m	300 mm for the height of one storey and 200 mm for the remainder
Exceeding 9 m and not exceeding 12 m	Not exceeding 9 m	300 mm for the height of one storey and 200 mm for the remainder
Exceeding 9 m and not exceeding 12 m	Exceeding 9 m	300 mm for the height of two storeys and 200 mm for the remainder

For cavity walls the thickness of the wall should be at least 250 mm or the thickness required for a solid wall under Rule 7, plus the width of the cavity, except in the case of the upper storey of a two-storey dwelling house where the inner skin may be 75 mm (giving an overall width of 225 mm) subject to certain special rules including a doubling of the normal number of wall ties, roof loads being carried partly by the outer skin, a strong mortar being used, and the wall being not more than 8 m long and 3 m high (5 m high to a gable).

For the purpose of Rule 7, the heights are measured from the base of the wall immediately above the foundation or footing and not from ground or finished floor levels.

Walls are regarded for this purpose as being divided into separate lengths by piers, buttresses, chimneys and buttressing walls. These separate lengths are measured centre-to-centre of the piers etc, provided that piers and buttressing walls themselves meet the necessary requirements and are fully bonded.

Fig. 4.2 shows part of an end-of-terrace Victorian cottage. Note that the brick cross-wall construction is separated from the flank wall on both levels by door openings so that little or no effective cross-wall bonding exists. The stud cross-wall is of no benefit to the stiffness of the flank-wall either. The flank-wall is 225 mm thick and divided into three lengths, these being 3 m, 5 m and 3 m respectively, this flank-wall being carried up to form a gable-end with an effective height exceeding 3.6 m.

If the chimney breasts are removed a single length of 11 m of wall exceeding 3.6 m high results which should be a minimum of 300 mm thick at ground floor level. Since this is not so, the removal of the chimney breasts is not recommended; they add essential stiffness to the flank-wall construction. If soft lime mortars are used, the removal of buttressing from the flank-wall may still be ill-advised, even if a check by reference to Rule 7 raises no technical objection, because the slenderness ratio by reference to CP 111 is excessive.

Slenderness ratios (SR) should be calculated in accordance with the formula laid down in CP 111. For two-storey residential buildings the Code recommends that the SR for walls built in cement mortars should not exceed 24, or 18 if built in the older type of lime mortars (12 if more than two storeys).

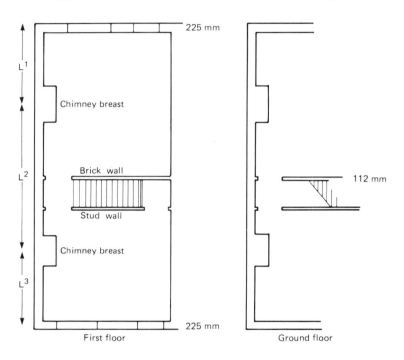

Figure 4.2 Ground and first floor plans of end-of-terrance Victorian cottage

The SR is determined by the length and height of the wall, the lateral restraint provided and the nature of the construction. A superficial examination of the overall construction will show if a building 'feels' right. Experience will tell whether a problem may arise having regard to the layout of the crosswall construction and age of the structure. Newer practitioners may find it necessary to check such details by approximate calculation until they have acquired the necessary experience.

Having assessed the condition of the main walls and the manner in which they are likely to cope with the forces imposed upon them, the surveyor should then consider the other factors determining whether those walls are likely to be satisfactory or not. These are: (a) their resistance to dampness rising up, penetrating through or percolating down the walls, (b) their thermal efficiency, and (c) in the case of separating walls (or party walls), their soundproofing properties. It is proposed to deal with each of these factors in turn.

Dampness

The building under review should be examined and compared with an equivalent modern building constructed to current standards. The extent that it falls short of modern standards should then be considered and the client advised accordingly. A good standard would be cavity-wall construction with clear, unobstructed cavities, having damp-proof courses at the base of walls at least 150 mm (two brick courses) above surrounding ground levels and with vertical and horizontal damp-proof courses at all points of contact between the outer and inner skin, and damp-proof cavity trays at all points of bridging. Additionally, there should be an impervious damp-proof membrane in solid floors contiguous with the damp-proof courses in all walls. By such means the passage of moisture into the interior from the exterior, or the site, is avoided.

The fact that a particular building is modern will not, of itself, ensure that a good standard is met since, in practice, poor detailing of cavity walls, obstructed cavities, punctured membranes and other faults abound. Nevertheless, a good yardstick to apply is that described. In so far as any construction under review fails to meet this standard, then this should be pointed out to the client, and many older buildings will not reach this standard.

Before about 1938, much residential construction was in solid brickwork or stonework and the main barrier against weather penetration in such construction is the condition and pointing of the external brick or stone face. In such circumstances this should be explained to the client and he should be advised to maintain the exterior surfaces in good order in future years and make a particular point of checking periodically for gutter and downpipe leaks. Leaking overflows will also cause considerable damage to internal plasterwork if neglected since the moisture can easily penetrate directly to the inside.

The clearance of 150 mm between exterior levels and damp-proof courses is recommended in order to prevent gutter-splash reaching the wall above damp-proof course and to allow some margin of clearance. Experience indicates that the ground levels around houses tend to rise by about 300 mm every 100 years, due to the addition of soil to gardens and the laying of new paving over old. It is essential to resist this trend.

When there is insufficient clearance to damp-proof courses it is recommended that ground levels be lowered. A general lowering of levels is best with the surplus soil or paving being carted away. Alternatively, a channel may be formed alongside the wall, at least 150 mm wide, suitably formed with cementwork sides and benching and having a fall-away from the main walls and adequate drainage to suitable storm gullies. A channel which fails to drain adequately is worse than useless since it will actually exacerbate any dampness. The brickwork then exposed at the base of the wall should be made good as necessary by pointing, cutting in new bricks or rendering to a suitable standard.

Vertical damp-proof courses of slates set into a rendered plinth are a traditional remedy. They are inserted between the high exterior levels and the walls where damp-proof course clearances are inadequate or the courses themselves are bridged. Such vertical courses often prove ineffective in the long run as the slates become porous and the bedding mortar between slates is rarely impervious to moisture. Any vertical slate damp-proof courses found between high exterior ground levels and the external walls of older buildings should be regarded with a critical eye and consideration given to a general lowering of levels or tanking applied to any at basement or semi-basement.

Damp-proof courses become porous with age and rising damp results. Specialist contractors now provide remedial services either by chemical injection of the walls or electro-osmosis to prevent capillary attraction in the walls, or by inserting new damp-proof courses in sections using a bricksaw to cut out chases in the walls, generally of about 1 m at a time. Electro-osmosis is now generally considered to be less effective compared with other methods and one large specialist firm has discontinued its use. Electro-osmotic damp-proof courses should be given special consideration if encountered during a survey.

Before allowing a client to employ such a specialist contractor it should be confirmed that the dampness found actually is due to rising damp. Many cases of diagnosed rising damp are not due to failure of the damp-proof courses at all but from penetration or bridging. It is significant that guarantees offered by specialist contractors will normally exclude liability for dampness due to causes other than moisture rising up the walls by capillary action. Before employing a specialist contractor it is recommended that obvious defects likely to cause dampness should first be eliminated and the structure subjected to a period of observation, if this is possible. It will often be found that dampness can be cured by such simple steps alone.

Having said this, occasions will undoubtedly arise when dampness may be diagnosed as rising damp due to failure of damp-proof courses. In such circumstances all main walls will need attention since the damp-proof courses will be of the same age and type unless parts of the structure date from different periods. The provision of new damp-proof courses in only a part of a building and leaving the original damp-proof courses in other parts may be unwise since further work in the other areas will be necessary in due course if the damp-proof courses are failing due to their type and age.

Specialist contractors are commercial men seeking work and there is nothing wrong with that. As a professional, however, the surveyor should be able to say whether or not the specialist's proposals are unnecessary or premature.

Bridging of damp-proof courses can arise from inside the building as well as from outside. Frequently there is a build-up of rubble and debris under suspended timber floors. Whenever such floors have to be exposed for building works or services it is recommended that the over-site areas be cleared out and swept and a check be made that all vents are free from blockage. Solid floors in older buildings may also bridge the damp-proof courses. The use of membranes in solid floors was not general practice in pre-1939 construction, and in halls, kitchens and sculleries quarry or ceramic tiles were often laid directly on the screed and thus provided a passage for moisture to rise up into the base of adjoining walls. It will often be found that plaster at the base of walls next to such floors has been affected by damp and a high moisture-meter reading obtained due to this damp transference.

In modern construction the failure to ensure that membranes are lapped over or under damp-proof courses and sealed at all joints can give rise to local bridging. Polythene can also tear or puncture, especially if laid under a concrete slab on gritty or stony blindings. Often, 500-grade polythene has been used as a cheaper substitute for the more durable 1200-grade which the writer would recommend for this purpose.

Thermoplastic tiles and woodblocks, commonly employed in houses of the nineteen-fifties, were laid in a bituminous backing which was itself the damp-proof membrane. Sometimes such tiles or blocks are removed by a do-it-yourself owner and replaced with other surfaces, the bituminous backing being disturbed or removed altogether. Modern vinyl asbestos tiles or welded sheet vinyl flooring cannot be laid with spot adhesive on a floor surface unless this surface rests over a suitable membrane. Failure to observe this will result in the floor screed sweating under flooring and causing it to lift and loosen. The bituminous backing to tiles and woodblocks should be contiguous with the damp-proof course in the walls but this may not always be so and can also be a source of bridging and moisture transference into the wall above damp-proof course level.

In older houses with damp solid floors or with timber floors in poor condition below recommended levels a useful remedy can be to have them replaced by new solid floors at a higher level. This can often be done usefully in kitchen and scullery areas where ceiling heights are sufficient. The ceiling height should not be reduced below 2.3 mm (the old Building Regulation minimum), preferably higher. The new floor can then incorporate a damp-proof membrane lapped at the sides and contiguous with the damp-proof courses in the walls. These should first be exposed by chasing the wall plaster and removing skirting boards as necessary. The client will then be able to lay vinyl asbestos tiles, sheet flooring or similar surfaces on the new screed in the knowledge that the floor will be dry and will not transfer moisture into the surrounding walls.

Apart from normal moisture-meter tests which should be made at selected points around the base of the main walls and at other points at risk, the wall surfaces should also be examined for signs of deterioration in the plaster, staining to wallpaper and mould growth, all signs of dampness (although not necessarily rising damp). Rust stains from skirting board nail heads are a good indicator of moisture behind the skirting board.

One final point regarding dampness in modern buildings is that in many post-1945 houses the pipework for plumbing and heating systems has been buried under solid floor screeds above the membranes and if an isolated point of dampness is found at the base of a wall, especially an internal wall, it is well worth while investigating the possibility of a leak in the system under the floor. If a serious leak is taking place from a heating system this can be confirmed by tying up the header tank ball valve, marking the side of the header tank at the existing level and then checking this level periodically to see if it drops. Any rapid loss of water from the header tank will indicate a leak to the central heating pipework in the floors which will have to be traced and remedied.

Thermal insulation

The thermal insulation requirements in current Building Regulations are reasonable standards to adopt and if the building under review is likely to be less satisfactory then the client should be warned. Generally, a cavity wall with insulation block inner skin and cavity fill is a minimum present-day requirement. Older solid wall construction will allow greater heat losses, as will all-brick cavity construction of the nineteen-thirties and nineteen-fifties, especially if the cavities are vented and open at the top as was common practice during those times. The standard of insulation of an all-brick cavity wall with well-ventilated cavity is, in practice, not much better than that of a half-brick wall.

Still considering thermal insulation, it should also be borne in mind that some buildings are inherently less efficient than others. Bungalows have twice the roof area for the same floor space as two-storey houses and a bungalow with a number of projecting wings and bays will be presenting a large external surface area, especially if it is detached. The ideal structural shape from this point of view is a square box which encloses the maximum volume for the least external wall area. In view of the importance of thermal insulation, this topic should be covered by the surveyor in his report and if any steps should be taken to improve insulation, then these should be incorporated in the report.

All brick cavity walls could be provided with sleeved vents through the cavities and the cavities sealed at the tops. Cavity-wall fillings may be employed. Solid walls can be improved by the provision of 3-mm polystyrene insulation under wallpaper on inside faces of exterior walls, and this will compensate to some extent for the lack of cavity construction. 'False' walls may be employed inside exterior walls with dry linings on battens and insulation.

Sound insulation

Current Building Regulations provide that party walls must be of 225-mm brickwork or dense blockwork, plastered both sides, to give the necessary sound-proofing. In practice, the older, denser party walls, often incorporating substantial chimney breasts, have been found to be more satisfactory in this respect than the more modern party walls. In a Building Research Establishment current paper, 'Field Measurement of the Sound Insulation of Plastered Solid Brick Walls' (BRE CP 37/78 March 1978), it was found that in houses built during the nineteen-seventies the test results indicated that walls constructed to meet the deemed-to-satisfy specification of the Building Regulations actually failed to meet that standard in practice in about 35% of cases.

The reasons for this include the fact that voids may arise in brickwork and blockwork if the workmanship is poor and frogs not fully filled. Chases for electrical and aerial conduits etc, often corresponding in location on both sides of a party wall, may allow local channels for sound to pass. In so far as it is possible to do so, the surveyor should comment upon the degree of sound insulation in separating walls, although in many cases this will not be possible.

Wall design

Fig. 4.3 shows a sketch section through an unrendered solid brick wall of a type commonly encountered in Victorian and Edwardian construction having a solid internal floor covered with quarry tiles and no effective damp-proof membrane. A risk of damp penetration through the wall and dampness rising by capillary action via bridging arises with this type of construction. Wet-rot or dry-rot to grounds and skirting boards in such a location is likely if dampness is present. Simple remedies such as dealing with defective rainwater equipment and lowering ground levels may suffice but it will not be possible to confirm whether the slate damp-proof courses are effective or not since they cannot function properly anyway until the obvious problems are dealt with. Floors may be re-surfaced with a new covering of two good coats of 'Synthaprufe' or similar bituminous coating, taking care to cut back old tiles, to expose the damp-proof courses and to finish the bituminous backing against the slates to ensure a continuous damp-proof course and membrane. Pointing and soft brickwork should also be attended to as required.

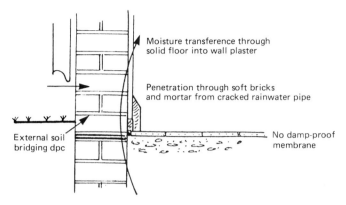

Figure 4.3 Bridging of slate damp-proof course in unrendered, solid brick wall

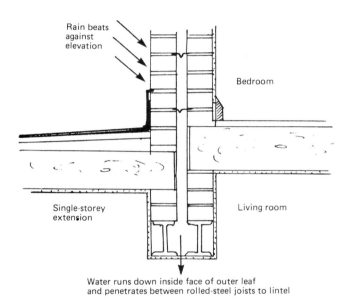

Figure 4.4 Lack of damp-proof cavity tray to cavity wall over opening for living-room extension

The standard of thermal insulation of such a wall will be poor and heat losses excessive by modern standards.

Fig. 4.4 shows a sketch section through a cavity wall where a living room has been extended and two small rolled-steel joists inserted. In such a location, especially in conditions of high exposure or normal exposure facing the prevailing weather (usually south west), a damp-proof cavity tray should be provided, lapped over the exterior roof flashing and a few open perpends or weepholes allowed for so that any water running down the inside face of the outer skin is picked up and taken to the outside over the roof line. Similar provision is required in a cavity wall where a subsidiary pitched roof joins the outer skin using a conventional flashing and where the cavity continues downwards to terminate at lintels over rooms below.

The provision of rolled steel joists and other lintel work in residential extensions of this type is often badly done but the detail can rarely be checked. Apart from the provision of proper damp-proof cavity trays and weeping perpend joints as necessary, it is essential that proper bearings for joists are provided and that for all but the very short spans, the rolled-steel sections should be bolted together with spacers to provide improved stiffness and resistance to torsion.

Occasionally, walls will be found which are rendered both sides and which will appear from measurement to be of brickwork. Care should be taken in making assumptions since such walls may be of either solid or cavity concrete construction, formed originally in sections using shuttering. There was a vogue for such construction for a time in certain localities in the inter-War period, and more recently the development of such construction using no-fines concrete has been popular. The nature of the construction of a rendered wall may often be confirmed by checking inside air vents and at the top of walls around the perimeter of the roof-space. Good building practice requires that wall cavities be closed at the top but in practice open cavities are often found, especially in older cavity walls. It is possible to look down such cavities from within the roofspace in some cases in order to check for blockages, bridged wall ties and other defects as well as to confirm the nature of the construction used for the outer and inner skins. Inspection within open wall cavities or other concealed areas can be undertaken using endoscopes or borescopes which require only narrow access through an airvent grill or a small drilled hole in one skin – endoscopes cannot be used to any advantage where the cavity is filled with insulation however. This equipment is not generally used as a routine test in all surveys but is particularly suitable for checking older cavity wall construction for corroded wall ties. Requiring a drilled hole about 10 mm in diameter, the endoscope can be used to inspect floor and roof voids, ducts and casings, is illuminated and can be linked to a 35 mm camera.

As regards cavity-wall construction, whether of brickwork, stonework, blockwork or a combination of materials, the overall stiffness of the construction relies upon the two skins being held together by effective wall ties. Since some cavity walls of a considerable age may now be encountered, many being more than 60 years old, it is worth considering how effective the wall ties are now likely to be.

Current Building Regulations require wall ties specified or manufactured to British Standard BS 1243:1978, 'Metal Ties for Cavity-wall Construction' or an equivalent including non-ferrous or stainless-steel wall ties. Galvanised wall ties usually take the form of either vertical twist, double triangle or butterfly types, and the durability of the ties is dependent upon the zinc coating given during galvanising. The possibility of corrosion and the possible resulting instability of cavity walls is a known hazard and some reports of structural failure of cavity walls for this reason have been made, especially in cases where black-ash mortar has been used. In a BRE information paper, 'The Performance of Cavity-wall Ties', (BRE Information Paper IP 4/81 April 1981), it was concluded that vertical-twist ties may lose their zinc coating within 23 to 46 years and wire ties in similar conditions within 13 to 26 years, assuming zinc coating to the current British Standards. The use of wire ties was considered unsatisfactory, and more durable ties were recommended although the majority of existing buildings were regarded as sufficiently robust for widespread problems to be unlikely.

Faced with cavity-wall construction and black-ash mortar (where an ash-fine aggregate has been used instead of sand) the surveyor should beware of corroded ties. Faced with any cavity-wall construction over 50 years old he should be on his guard against the possibility that wall ties may have corroded to the point where there is no effective connection between the two wall skins. Conventional low-rise cavity walls will normally remain quite stable, even in the absence of effective ties, due to stiffness afforded by bonded cross wall construction inside and abutments over, under and around openings, these abutments having only limited resistance to tension between skins but good resistance against compression. Corrosion of ties could be more significant in low-rise construction with little bracing and in high-rise buildings which have masonry infill panels with galvanised metal ties.

The possibility should be borne in mind that galvanising of connecting ties used for certain types of pre-fabricated construction may be poor. Surveyors will occasionally have to report on local authority houses (some now privately owned having been purchased by the tenants) constructed by using various forms of pre-cast concrete slabs and pre-fabricated panels. A number of different systems were used during the post-War municipal housing programmes. Variations in such construction will also be encountered in commercial and industrial structures including pre-cast concrete panels fixed using bolted connections and spacers with galvanising to bolts and fittings often of a poor standard. In addition to problems associated with rusting of galvanised fixings, some systems incorporated a framework of tubular supports which are prone to corrosion and the pre-cast concrete panels themselves may be subject to the rusting of the reinforcement bars inside, this causing disintegration of the panels.

Prompted by reports of deterioration to prefabricated reinforced concrete houses the Building Research Establishment has published reports on individual PRC house types including Boot, Cornish, Unit, Orlit, Unity, Wates, Wollaway,

Figure 4.5 Large panel pre-cast reinforced concrete construction from the 1960s

Figure 4.6 A typical bulge fracture, wide in the centre and narrowing to hairline at both ends

Reema, Parkinson Framed, Stent, Dorran, Myton, Newland, Tarran, Underdown, Winget, Ayrshire County Council (Lindsay) and Whitson-Fairhurst construction. Findings are summarised in Information Papers IP 16/83· and IP 10/84 available from BRE. As part of an on-going programme the BRE intend to report on all the main types in time.

The main problem is corrosion of reinforcement which expands causing concrete to crack and spall and investigations revealed that the reinforced concrete components are gradually deteriorating. Whilst the great majority of houses studied were found to be in structurally sound condition there was a wide range in the rates of deterioration both between and within types and there are no practicable techniques which can halt deterioration altogether. It was advised that houses of this type be regularly inspected and adequately maintained to prolong their remaining life by measures such as are described in BRE Digest 265, 'The repair of reinforced concrete'.

Monitoring walls

Occasionally walls will be either quite new or have surfaces recently covered in such a way that evidence of past damage will not be discernible. In these cases procedures will be available for monitoring the performance of the walls and sometimes, in an initial report, it may be necessary to recommend such procedures.

One of these recommended procedures is the accurate monitoring of crack damage in distressed buildings, this generally being undertaken using calibrated tell-tales, these being rigid indicators fixed over cracks walls and showing measured horizontal and vertical movements over a period of time. Tell-tales should be given reference numbers and all measurements of movement logged at regular intervals in a permanent log book. After a suitable period has elapsed, the information collected will be useful in showing the extent and direction of structural movement and whether or not it is seasonally variable.

Movements in walls due to thermal expansion and contraction can show a distinct pattern of measurement, with conditions in hot and cold weather differing. Movement in walls due to progressive settlement of parts of a structure relative to other parts will show a general progressive overall measurement. Structural settlements followed by heave due to shrinkage and swelling of clay soils will produce distinct seasonal patterns. Expansion due to chemical deterioration of concrete members may be found to be more noticeable during periods of wet weather. The variety of movements which may be caused by subsidences and landslips of the subsoil are considerable, and by accurate monitoring over a period of time it is possible to confirm overall downward movements of parts of a structure, progressive leaning and tilting of walls, rotational movements of projecting wings and other structural manifestations.

Interpretation of crack monitoring results should be undertaken with care since it is sometimes easy to jump to wrong conclusions. Movements on shrinkable clays are particularly tricky to interpret since what appears to be a downward settlement of part of a building could in fact be stability with the other part moving up due to clay swell. The use of an external datum may be needed to confirm the point.

Clearly not all clients will be in a position to benefit from the monitoring of crack damage since if the client is a prospective purchaser he may need to decide quickly whether or not to purchase a building. If he is proposing to take a repairing lease he may have no time to spare for lengthy structural monitoring. Similarly, prospective mortgagees will be unlikely themselves to be able to commission structural monitorings of walls prior to confirming any mortgage advance although they may be able to require the prospective mortgagor to undertake this and allow a suitable period for him so to do prior to confirming their offer. This situation is unlikely to be helpful to a prospective mortgagor who is himself an intending purchaser and under pressure to exchange binding contracts.

One feature of crack damage occasionally encountered in Victorian and Edwardian construction of a cheaper type is that some solid walls were built without cross-bonding and snapped headers used to the outer face, such walls having the appearance of 225-mm solid brickwork but, in fact, comprising two skins of 112-mm brick with a continuous lime mortar joint rising vertically in the centre. Curious crack patterns and leaning and bulging of such walls may be found showing different patterns to inner and outer faces. This construction was used to minimise cost by saving all facing bricks for outside and in practice may be found to develop instability with age, especially where tall gable-ends are concerned.

Retaining and freestanding walls

In addition to the main walls of buildings, surveyors often have to assess exterior walls such as retaining walls, freestanding walls or parapet walls (generally of brickwork or stonework) which form adjuncts to or extensions of the main structure. Unfortunately such walls are frequently badly built due to lack of understanding of the principles involved and (in most cases) the fact that they are outside Building Control Legislation.

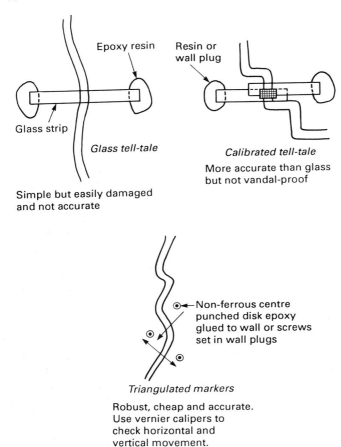

Epoxy resin

Resin or
wall plug

Glass strip

Glass tell-tale

Simple but easily damaged
and not accurate

Calibrated tell-tale

More accurate than glass
but not vandal-proof

Non-ferrous centre
punched disk epoxy
glued to wall or screws
set in wall plugs

Triangulated markers

Robust, cheap and accurate.
Use vernier calipers to
check horizontal and
vertical movement.

Figure 4.7 Crack monitoring

Such walls are examined to see whether or not they are plumb and level. The overall construction should be considered in relation to the importance of the walls and the requirements of their construction. Clearly, retaining walls may be very important indeed to a structure on a sloping site and may be essential to the stability of that site and thus of the structure.

The first design consideration is whether or not the walls under review are allowed adequate provision for movement. Generally, movement joints in brick retaining and free-standing walls are desirable at about 10-m intervals to cope with normal thermal expansion and contraction and occasional movements which occur as a result of chemical conversion in concretes and mortars. Movement joints at greater frequency may be recommended for free-standing walls on sites where the subsoil may be unstable and where, for reasons of economy, a shallow foundation is necessary. On shrinkable clays, for example, garden walls and the like could be provided with movement joints at 3- to 5-m centres, obviating later cracking in the brickwork due to seasonal ground movements.

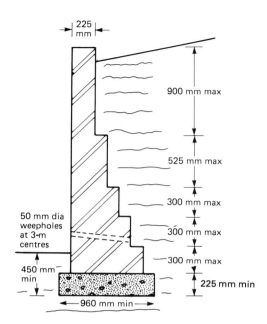

225 mm

900 mm max

525 mm max

300 mm max

300 mm max

300 mm max

225 mm min

50 mm dia
weepholes
at 3-m
centres

450 mm
min

960 mm min

Figure 4.8 Typical design for all-brick
retaining wall, using special-quality
brickwork which does not require tanking
or weathering. The wall is stepped back
in half-brick increments. Backward-
sloping drain-holes prevent staining down
the face of the wall from very small
quantities of groundwater discharging
over a long period. Note that ordinary-
quality brickwork requires special
tanking, damp-proof courses and
weathering, otherwise frost damage will
result

The second design consideration is whether in the case of retaining walls the construction itself is adequate. Where a retaining wall is essential to the stability of a structure a consulting engineer's report should be obtained if the construction standard is doubtful. This will generally involve some site investigation behind the wall. In *Fig. 4.10* a cross-section of a brick retaining wall is shown. It will be noted that 225-mm brickwork should be used only for retained heights up to 900 mm, and that progressively thicker wall construction is required as the height of the wall increases. Regard should be given to the bearing pressure values of the retained ground, which may require special consideration. *Fig. 4.9* shows a typical free-standing wall. It should be noted that 225-mm brickwork is a minimum thickness

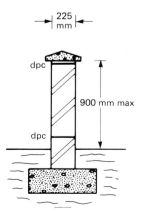

225 mm

dpc

900 mm max

dpc

Figure 4.9 Freestanding wall construction. Throated
weathering projecting at least 45 mm, plus damp-proof courses
at top and bottom, allow ordinary-quality brickwork to be
used for normal exposure conditions, but special-quality
bricks are necessary for footings below dpc

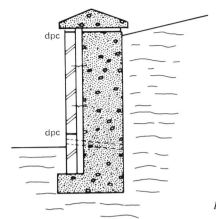

Figure 4.10 Mass concrete retaining wall with brick face

and that piers or buttresses or, alternatively, thicker overall construction is required for heights greater than 900 mm. A check on the slenderness ratio is recommended in doubtful cases, paying particular regard to the strength of the mortar, which is often poor, exterior brickwork and stonework, especially if this is old.

The third design consideration is whether the wall material will deteriorate should groundwater be able to enter the wall, and taking into account conditions of exposure. Special-quality brickwork or stonework resistant to frost damage and chemical attack from groundwater will be normally be suitable. Ordinary-quality brickwork or stonework which may not resist such deterioration should not be used in conditions of high exposure. In the case of normal exposure, ordinary-quality brickwork or stonework may be used provided that certain precautions are taken, these being the provision of tanking to retaining walls, and horizontal damp-proof courses at the tops and bottoms of walls, also the provision of adequate copings with throated over-hangs. In *Fig. 4.10* a sketch of a mass concrete retaining wall with a facing of ordinary-quality brickwork is shown. Below damp-proof course at the base a special-quality brick footing is necessary. In *Fig. 4.9* and *4.10* the damp-proof course is shown as a line. For most situations however a simple bituminous felt or plastics damp-proof course should not be used due to the weak bonding line which results and the risk of overturning failure of the wall – a better solution is to use two courses of engineering brick to provide a damp-proof course with uninterrupted mortar joints within the construction.

Chapter 5

Floors

When we first bought the house we thought the ramps in the floors at the door openings were there because the house was built for someone with a wheelchair. It was only later that we realised that the floors have all sunk by three or four inches, which is why a second set of skirting boards has been fixed lower down on the first set of skirting boards.
Extract from letter from potential client to surveyor

During the survey the nature of the floor construction in each part of the building should be confirmed as far as possible in the circumstances.

The ground floor may be of suspended timber construction, or suspended concrete slab construction, or solid slab construction on a hardcore base. Settlement problems in solid slab floors have been found by the National House-Building Council (NHBC) to be a common source of complaint in new houses and they now require that any construction on sloping sites which needs deep fill under solid floors should be built using suspended floors instead, as detailed later in this Chapter.

Suspended timber floors

Suspended timber ground floors may be of the traditional type supported on intermediate sleeper walls or may be of the fully-spanned type. With either type it is clearly essential that the decking and the supporting timbers are of adequate type and size for the spans involved and a check by reference to the standards laid down in Building Regulations is recommended, using the appropriate approved document. The old 1976 Regulations contained a helpful table in Schedule 6 and this is reproduced below since it provides a useful standard to apply to older residential buildings.

For Surveyors still familiar with imperial measurements, a rough rule of thumb for use in old buildings is that the depth of the joist in inches should be not less than the half the span in feet plus one. So for a span of 12 feet: 7" × 2" joists at 16" centres would be a minimum for residential loading. For a span of 10 feet: 6" × 2". For a span of 14 feet: 8" × 2". For a span of 16 feet: 9" × 3" (joist width should be increased so as not to be less than a quarter of the depth).

In older houses the surveyor may encounter uneven and sloping floors, some
with considerable gradients when floor joists are over-spanned, either being so
originally or having had intermediate support removed from below. An overall
gradient in one direction will, of course, not be due to deflection but will indicate
some more fundamental movement in the bearings to the floor joists. Faced with a

*Schedule 6, Table 1, 1976 Building Regulations (applicable from 1976 to 1985 – now
replaced by Approved Documents under current Regulations)*

Floor joists for dead loads of not more than 25 kg/m^2 excluding the mass of
the joist.

Size of schedule joint (mm)	Maximum span to centres of bearings (m)		
	400 mm spacing	450 mm spacing	600 mm spacing
38 × 75	1.03	0.93	0.71
38 × 100	1.74	1.57	1.21
38 × 125	2.50	2.31	1.81
38 × 150	2.99	2.83	2.46
38 × 175	3.48	3.29	2.86
38 × 200	3.96	3.75	3.26
38 × 225	4.44	4.20	3.66
44 × 75	1.18	1.06	0.81
44 × 100	1.97	1.78	1.38
44 × 125	2.62	2.52	2.05
44 × 150	3.13	3.02	2.64
44 × 175	3.65	3.51	3.07
44 × 200	4.16	4.50	3.93
50 × 75	1.33	1.19	0.92
50 × 100	2.10	1.99	1.55
50 × 125	2.73	2.63	2.29
50 × 150	3.26	3.14	2.81
50 × 175	3.80	3.66	3.27
50 × 200	4.33	4.17	3.72
50 × 225	4.85	4.68	4.18
63 × 150	3.51	3.38	3.09
63 × 175	4.08	3.93	3.59
63 × 200	4.65	4.48	4.10
63 × 225	5.21	5.03	4.60
75 × 200	4.90	4.93	4.33
75 × 225	5.49	5.30	4.85

floor which slopes in one direction only, the surveyor should confirm whether or not this is due to failure of the joist bearings or to a downward settlement of the wall on which they bear. A wall plate, for example, which is eaten away by beetle attack or softened by rot may simply crush under the weight of the joist ends resting on it and a downward movement equal to the depth of the wall plate could occur. This is a particular danger in old structures with wall plates built into the damp wall construction.

Upper floors in traditional dwelling-house construction are normally suspended timber floors, fully spanned between ground floor walls and supporting certain point-loads such as airing cupboards and tanks, blockwork or stud partitions and the like. A check should be made on such floors similar to that recommended for ground floors, especially having regard to the possibility of over-spanned joists in older buildings.

For bearings greater than 100 mm a notional bearing of 100 mm may be used so that in most cases the maximum clear span will be 100 mm less than the span to bearing centres shown above.

The Tables to the Building Regulations assume new timbers in good condition, classified according to the current British Standard, or graded in accordance with certain minimum standards for a particular species. It will be appreciated that timbers in existing buildings, and especially older buildings, may not meet the grades of Group 11 specification laid down and some allowance for this may need to be made in applying Table 1 as a rule of thumb to check the adequacy or otherwise of floor joists, particularly where the span is marginal.

In domestic construction the most commonly found joists will be 50 mm × 150 mm, 50 mm × 175 mm and 50 mm × 200 mm, generally at 400-mm centres, with maximum permissible clear spans of 3.16, 3.7 and 4.23 m respectively (i.e. the Table spans less the distance to bearing centres on notional 100-mm bearings).

In domestic construction which has suspended timber ground floors resting on intermediate sleeper walls, the joists will normally be found to be 100 mm × 50 mm and the requirements of the old Building Regulations would assume clear spans between sleeper walls of 2 m, assuming 400-mm spacing and dead loads not exceeding 25 kg/m^2. It follows that ground floor construction with spans between sleeper walls exceeding 2 m clear may be over-spanned and this is a useful avenue of investigation for the surveyor to explore if the floors appear to flex unduly underfoot. In some newer properties where joist sizes have been calculated to comply with the Building Regulations rather than use of Tables, such unacceptable flexing may often occur and can prove embarrassing, particularly, for example, in dining rooms where walking across the floor can make cutlery rattle.

Intermediate strutting to suspended timber floors is required where the span exceeds 2.5 m, using herringbone or other strutting to prevent undue warping or torsional distortion of the joists. The *Registered House-Builders' Handbook* recommends and requires one set of such strutting for spans of between 2.5 and 4.5 m and additional strutting for greater spans. It is important that the strutting be neatly done and tightly fixed otherwise it will not prove effective. In the event of strutting being absent in spans over 2.5 m, the surveyor may have to consider whether or not any undue warping or distortion of the joists has resulted, and recommendations to this effect are required. In practice, intermediate strutting is often found to have been omitted with no apparent ill effects.

It is important that bearings and floor joists be so constructed as to eliminate the possibility of moisture transference from adjoining masonry and from oversite into timbers. In older buildings, joists are often built into the outer walls where the construction is in solid brickwork and a likelihood of rot attack to joist ends then arises, especially if the wall concerned faces the prevailing weather or is in a porous condition (see *Fig. 21.1*).

In some older buildings the joists will rest on wall plates which are themselves built into the outer walls or fixed to the face of walls, and the possibility of decay and even complete disintegration of the wall plate arises, causing serious loss of bearing to the joists.

It may also be found that sleeper walls or other means of support lack effective damp-proof courses under sleeper plates and joists, thus giving rise to decay and insect attack to the timbers. Commonly the air-space under a suspended timber ground floor will be found to contain large quantities of debris left over from the original construction or later building works, and this can cause moisture transference into timbers from the oversite and also block ventilation.

The provision of adequate underfloor ventilation to a suspended timber floor which is above an oversite in contact with ground moisture is essential since lack of good ventilation can give rise to condensation on the timbers and resultant decay. A careful check must be made for any 'dead' areas of air-space under such floors which may not be adequately ventilated. Generally a good standard of underfloor ventilation to a suspended timber ground floor in domestic construction would be 225 mm × 75 mm terra cotta airbricks, sleeved through wall cavities, sited at 1.5-mm centres. Iron grilles are not recommended since eventually they rust and permit the access of vermin. A vent should be provided at or near each extremity of the floor void to prevent areas of 'dead' air.

Solid floors

Solid slab ground floors of concrete on hardcore are more durable in the long term than suspended timber ground floors. In older buildings where timber floors are in poor condition, replacement with a solid concrete floor is often to be recommended for a number of reasons, especially if the adjoining ground levels are high and it is difficult to provide the necessary clearances to air vents.

The main problems likely to arise with solid floors at ground-floor level will be due either to settlement of the floor slab or rising dampness. Settlement and dishing of post-War solid concrete floors due to inadequate consolidation of the hardcore filling beneath or the use of unsuitable fill material is a common problem, so common in fact that the NHBC list it as being the major source of complaint in houses covered by their scheme.

On reasonably level sites, the NHBC will still approve solid concrete ground floors on a hardcore base and, in theory at least, such construction should be quite suitable for most ground conditions provided that the standard of worksmanship is satisfactory. Most problems arise when workmanship is poor and site supervision lacking.

On sloping sites where a deep fill would be required under concrete ground floors, the NHBC now require suspended-floor construction to be used where fill would exceed 600 mm in depth. On a sloping site it may be that a mixture of solid concrete on fill and suspended timber (or concrete) floors would be necessary to comply with this requirement (NHBC Practice Note 6, 'Suspended-floor Construction for Dwellings on Deep-fill Sites').

Unless the laying of fill material is supervised constantly and thoroughly it is virtually impossible to ensure that consolidation is adequate and that all material used is suitable.

Occasionally problems may arise due to the use of fill material which is still chemically active, such as cases where industrial waste material or ash blinding is used. It is essential that all fill material be chemically inert, otherwise damage to floor slabs from expansion or contraction of the fill are possible and chemical attack to the concrete and surrounding masonry could occur.

It is also possible for a solid concrete ground floor slab to be built over ground affected by existing leaking drains or for such drain leaks to arise after construction. Occasionally old land drains may exist and continue to discharge groundwater into the subsoil under the slab, thus causing erosion.

When inspecting a building with a solid concrete ground floor the general floor levels and amount of deflection which may be present in relation to the age of the floor should be considered. Some initial consolidation is often inevitable. If the floor has been in place for some years and the extent of any deflection is acceptable (say up to 0.003 of the span), this may be regarded as a matter for mere comment. If the deflection is greater then it may be necessary to consider undertaking investigations under the slab to confirm the nature and quality of the fill material. A core sample or small sample inspection hole is the best means of achieving this result. The sample material should be examined and, if necessary, a laboratory report obtained. When taking samples care must be taken to check for excessive moisture under the floor. Under normal conditions both the fill material and any subsoil under a floor should be nearly dry. Excessive amounts of water in samples may indicate a plumbing or drain leak under the floor and this could be one cause of the problem.

Where solid concrete floor slabs have settled due to inadequate consolidation of the original fill, a remedy may be required to stiffen the fill. This can be carried out, without taking up the floor, by using pressure-grouting techniques. Floor surfaces can then be re-screeded level, taking care to ensure the integrity of damp-proof membranes and also checking that membranes remain contiguous with the damp-proof courses in the surrounding walls in order to prevent moisture transference.

An examination around the perimeter of a floor is always to be recommended since this could reveal any gaps which may have opened up below skirting boards. If the surveyor is satisfied that the skirting boards are at their original height, a measurement of the gaps will indicate the extent of any downward movement around that perimeter. If quarter-round beading or similar additional trim has been applied at the base of skirtings, this may indicate that gaps have appeared and subsequently been covered. This feature should place the surveyor on this guard over the possibility of settlement in that floor. New skirting boards may have been applied and this could be due to a number of causes, including the possibility of such settlement in the floor surface.

Floors should be checked for level using a spirit level where a visual inspection suggests a slope. Some surveyors even carry a marble and use this to see if solid floor surfaces are uneven. However, since most floors are uneven to a greater or lesser degree, the fact that a marble may roll on the floor will not of itself be a means of accurately gauging the extent of such unevenness. A long spirit level is much better. In a larger building it may be necessary to take levels using a dumpy level or water level in order to assess the degree of unevenness in the floors.

A series of spot levels will often reveal considerable fluctuations in a wide-based structure which are not apparent visually.

In good construction, timber and suspended concrete ground floors will be separated physically from any contact with the site or structure below damp-proof course line, and floor surfaces should therefore be dry. Solid concrete ground floors are made dry by the provision of a suitable damp-proof membrane, and in the absence of an effective membrane the surfaces of such floors become damp due to moisture rising up from the site, especially if the surfaces are covered with carpets or sheet flooring.

Damp-proof membranes may be located below the floor slab, in which case polythene film of at least 1000 gauge, and preferably 1200 gauge, should have been used, or alternatively, a bitumen sheet to British Standard 743. It is essential that all joints in such membranes be lapped and sealed and that the membranes are carried up around the perimeter and lapped under the damp-proof courses in the surrounding walls so as to provide a completely imperforate barrier against rising moisture. Lapping should be at least 300 mm. Such membranes should have been laid over a suitable blinding of inert ash or sand, free from any sharp protrusions which could puncture the membrane.

Damp-proof membranes in sandwich construction may be located within the slab and could comprise polythene film or bitumen sheet (as before described), hot-applied mastic asphalt, pitch or bitumen, 0.6 mm minimum thickness of cold-applied bituminous solutions, or composite polythene and bitumen self-adhesive. In such construction the concrete slab is first allowed to dry thoroughly and the membrane applied to the surface and lapped up to the damp-proof courses as necessary. A surface screed is then applied which, to avoid 'shelling' or lamination, should be a minimum of 50 mm depth.

Membranes also may be located on the floor surface in the form of hot applied mastic asphalt, pitch mastic or cold-applied pitch/epoxy resins. In older construction hot-applied mastic will generally be found as a base for woodblocks or thermoplastic tiles, the use of polythene and other modern materials not becoming general practice until post-1960.

In pre-1939 and older construction, solid floors were often not provided with any damp-proof membranes at all and the surveyor will often encounter flagstoned floors, quarry-tile surfaces or ceramic tile surfaces of various types and ages which may be performing adequately when fully exposed but which would immediately begin to 'sweat' if covered with tiles, sheet flooring or carpeting. In advising a client in such circumstances, and before modern surface materials may be applied, it is essential to explain the significance of the modern practice of providing membranes and the special attention required to older floors which lack such membranes.

For the first half of this century most housewives were content with a kitchen equipped solely with a sink and draining board, gas-point for an oven and a quarry-tile floor which could be periodically washed down with soap and water and polished on Sundays. Quarry tiles are not impervious to moisture but they present a reasonably adequate barrier and the need for a dry floor surface was not important in those days. Nowadays a wife would wish to see a modern kitchen installed in an older house in order to make it acceptable and the old quarry tiles replaced by vinyl asbestos tiles or sheet vinyl floorings. It is essential that any new floor surfaces be laid over a suitable membrane which is continuous with the damp-proof courses in the surrounding walls. To achieve this it may be necessary to remove skirting boards together with wall and floor material at the skirtings in order to expose damp-proof courses in the walls and to seal the new membranes against these damp-proof courses to avoid moisture transference.

Flagstone floors may be lifted and re-laid over a suitable membrane if flagstones are required as a finish, again paying particular attention to the need to joint the membrane to the surrounding damp-proof course.

In commercial structures, solid floors may be laid as dry floors or without any membranes. Again, provided that the floor surfaces are exposed, it will often be found that concrete floors will remain reasonably dry on the surface even if no membrane is provided, but once the surfaces are covered they will sweat. The needs of the client may require consideration since certain types of storage and warehousing, for example, would require dry conditions and dry floor surfaces.

Concrete floors in commercial structures can be dusty and a suitable treatment of concrete paints or special sealants to reduce surface dusting may be recommended. For many types of manufacture a dust-free atmosphere is essential especially where precision work or electronics are concerned.

Where a garage is either an integral part of a house or is attached to it with a communicating doorway into the main accommodation, the garage floor will normally be a solid concrete slab. This slab should be on a lower level than the main ground floors inside the house with a step up at any doorway of at least 150 mm. The purpose of this is to prevent burning petrol in a garage from spreading under the door and into the main house. Garage floors should also have a slight slope down towards the main entrance to allow water to run away.

Particular attention should be given to providing adequate drainage to paved areas outside garages, especially where driveways slope down towards the building, in order to prevent driving rain or stormwater running in through the main door opening. Generally a driveway which slopes down towards a garage should always be provided with a suitable stormwater gully and a suitable drainage fall to this gully. This is often found, in practice, not to be the case with the result that ponding of stormwater in driveways and garages occurs. In such cases a channel and grating across the threshold of the door and drained into a gully should be provided.

A surveyor should confirm, if possible, the nature of the floor material under any fitted coverings. If fitted coverings are not to be lifted this should be made clear at the outset and a statement that the condition of floor surfaces cannot be confirmed included in the report. The nature of the floor surfaces should be suited to the purposes for which they are required and if this is not so then some advice on the point should be given.

Suspended concrete floors

Suspended concrete floor construction can take one of two possible forms, either pre-cast construction with or without a composite screed, or cast-*in-situ* construction. The latter is more commonly encountered in older buildings.

Cast-*in-situ* construction may be identified where the underside is exposed, mainly in commercial buildings, as the pattern left by shuttering is visible where it used to support the floor during curing.

When the floor construction is covered on all surfaces it will be impossible to confirm the nature of the construction without some special exposure or core samples being taken.

Pre-cast concrete floors appear in a wide variety of forms and slab sizes. It is more likely that defects due to conversion of concretes may arise in pre-cast construction, particularly if high-alumina cement (HAC) is present. HAC had an advantage in the production of pre-cast concrete members as it allowed a rapid drying time and quicker production. During the post-War period until 1974 a great deal of pre-cast floor slabwork and beams were made using HAC.

In February 1974 two roof beams in a school in Stepney collapsed and a major question arose as to the condition of beams and slabs in post-War buildings where HAC had been used. The Building Research Establishment undertook urgent investigations to establish the degree of risk likely to arise in buildings with pre-cast concrete beams made with HAC. The results of their investigations were published in April 1975 in Current Paper CP 34/75 'High-alumina Cement Concrete in Buildings'.

The outcome, according to the BRE, was that the risk of structural failure to floors of up to a 5-m span made with pre-stressed pre-cast beams using HAC is very small.

Spans of more than 5 m are unlikely to be encountered in residential construction, including blocks of flats, except in exceptional cases, bearing in mind that cross-wall construction is generally used and the spans will normally be the shorter of any two room dimensions. In most cases a surveyor will be able to see when the existence of spans greater than 5 m is unlikely in floor slabs, although it is rather more common in blocks of flats where beams are used. In commercial buildings spans exceeding 5 m will often be encountered in both floor slabs and beams. If the building was erected between 1925 (when the use of HAC in the United Kingdom began) and 1975 (when it ceased) the possibility of chemical conversion of the concrete resulting in a loss of bearing capacity may then arise.

In cases where there is a risk of the presence of HAC in spans exceeding 5 m, the surveyor should recommend that a consulting engineer be asked to carry out tests and advise independently on the position.

Floors generally

Floors should therefore be reasonably level, capable of taking dead loads and superimposed loads without excessive deformation and should be dry. Floors may also have to accommodate service ducting in commercial buildings and for such purposes a floor with accessible ducting to accommodate pipework and cables is desirable. In dwelling houses it is preferable that central heating pipework and other services be laid so that easy access is possible – in solid floors ducts in screeds (above the damp-proof membrane) are preferred and in timber floors screwed traps at important access points are useful. It is also a good idea to incorporate dovetailed perimeter battens in screeded concrete floors (above the damp-proof membrane) to accommodate carpet fixings for fitted carpets.

Chapter 6

Roofs and chimneys

I looked into the loft myself and thought it was rather light in there – I assumed there was a skylight. I now realise it was because so many of the slates were missing.
Note from client to surveyor when paying his account

There are two basic types of roof. Pitched roofs with an inclination of 10° or more and flat roofs with an inclination of less than 10°. The results of roof failures can be serious since leakage of water into a building may give rise not only to the need for repair of the defect but also consequential losses. Special attention should be paid to the requirements of clients interested in non-residential buildings such as factory and warehouse space. If materials are to be stored in circumstances where roof defects could cause damage, particular attention should be given to this point.

Life of roof coverings

Most roof coverings have a reasonable and well-defined effective life span beyond which re-covering may be necessary or trouble may be anticipated. In practice, there is considerable variation in the quality of materials from different sources, especially traditional materials, but the following is a list of the principal roof-covering materials likely to be encountered and a guide to their life span.

	Years
Hot-bonded, built-up mineral felt flat roofs to BS 747, well constructed	10–15
Hot-bonded, built-up felt roofs constructed to lesser standards	3–10
Good quality flat lead roofs, well detailed and laid with rolls	80–100
Good quality flat copper roofs, well detailed	30–75
Zinc roofs, well laid with rolls, in areas not subject to chemical pollution	20–40
Zinc roofs in situations of chemical pollution	15–20
Asphalt, well laid and detailed, in two coats to BS CP 144, Part 4:1970	20–40
Hand-made clay tiles with battens and sarking felt	60–70
Machine-cut clay tiles with battens and sarking felt	40–60
Plain concrete tiles, double lap with battens and sarking felt	50–90
Interlocking concrete tiles with battens and sarking felt	50–80
Interlocking Dutch or Belgian clay tiles	40–60

	Years
Asbestos-cement tiles or roofing sheets	20–25
Good quality Norfolk reed thatch	50–60
Straw thatch	20–25
Local stone slabs (depending upon porosity)	10–100
Good quality Welsh slates with battens and sarking felt	60–90
Poorer slates, subject to lamination with age	40–60
Shingles (depending upon quality and frequency of treatment)	5–30
Perspex, clear or obscure uPVC sheets etc. (depending upon brittleness etc.)	5–25

The surveyor should confirm during his survey the nature of the roof coverings in various parts and assess their likely age. He should then consider what advice may be given in each particular circumstance. Where roof coverings are likely to give problems in the foreseeable future the position should be explained to the client having regard to the possibility of either repair or renewal of these coverings. There will be many situations where re-covering might be recommended, but where the client could, if he so wished, extend the life of the roof more or less indefinitely by undertaking periodic overhauls and renewing individual defective slates or tiles as the necessity arises. Many clients are quite happy with an old roof covering which needs such periodic attention and will happily spend the money needed to renew it on something else. It seems quite reasonable in the circumstances for the surveyor merely to present the facts and leave it to the client to make the decision. Many examples of old slate and tile roofs will be encountered, still giving service well beyond the time scales indicated above.

Pitched roofs

Undersides of roof slopes will generally be lined with sarking felt in buildings erected post-1938 and in some cases earlier. Only a very limited view may be available of the underside of the roof covering if felt is loose or torn, and there will be no view of it at all if it is tightly and correctly laid. Unfelted roofs will normally allow an examination to be made of the underside of the covering, but where feather-edged boarding is used then very little access to the underside of the covering will exist. This is encountered mainly in 1920–1940 property.

It is helpful to examine debris which may have accumulated on ceilings in older roofs. Extensive deposits of clay-tile laminations and broken tile nibs will indicate that the tiles are laminating and becoming porous. Slate laminations, sections of cracked slates and rusted nail heads will all indicate that a slate roof is approaching the end of its recommended life span, or that 'nail sickness' is causing otherwise sound slates to slip. Often a roof covering will have been replaced and the debris in the roof void will indicate the type of previous covering, a point which may be significant if a heavy covering has replaced a lighter one since roof timbers designed for the lighter covering could deflect under the weight of the heavier.

Pitched roofs without underfelt should be regarded nowadays as below the acceptable standard. Clients should be warned that the degree of weather-proofing, especially in driving rain and snow, will not be as good as with an underfelted roof and also that the roof void will become dusty and items stored there will therefore need some protection.

The scantling of the roof timbers should be checked. In older buildings, rafters may lack effective purlin support, or purlins may lack sufficient strutting. Sagging rafters or purlins should generally not be jacked back into line but recommendations to stiffen up the purlin or strut construction to a reasonable standard should be given. A cross-check by reference to Schedule 6 of the Building Regulations should be made, bearing in mind that the older timbers may not be of an equivalent standard, and it is better to over-size purlins, struts and the like in marginal cases.

Figure 6.1 Old slate roofs can have their lives extended to some extent by fixing slipped slates with tingles

In older roofs a check for roof spread is advisable. The rafters should meet the ridge board tightly and if they have pulled away, this could indicate insufficient provision of collars, inadequate fixing of joists at the rafter feet, or insufficient binders to roof slopes which are at right angles to the main rafters.

In modern trussed-rafter and roof-truss systems certain important points are often overlooked. First, diagonal restraint is required to trussed roofs with gable-ends to prevent a parallelogrammatic, or 'racking', sideways movement of the trusses. Diagonal planking from the eaves at each corner to the ridge at the centre with two good galvanised nail fixings to each truss is essential, especially if the gable-ends themselves are of poor construction. Secondly, the end two or three trusses should be strapped with galvanised-steel strapping built into the gable-ends to prevent separation. Thirdly, no cutting or alteration of such truss construction should be undertaken, and if this is found to have been done, a warning should be given that the overall stability of the roof construction could have been seriously affected. Fourthly, only light storage on roof truss ties is advisable. The size and spacing of trussed rafters are not designed to accommodate heavy point loads.

Flat roofs

Most modern flat roof construction used in post-War dwellings is of hot-bonded, built-up mineral felt on timber decking, supported by timber joists and with a ceiling beneath. Built-up felt construction should normally comply with British Standard 747:1977 which sets out certain minimum standards for the materials to be used and the manner of laying. From a superficial inspection it is generally not possible to confirm whether or not the standards have been met. However, there are certain important matters which can be confirmed and borne in mind for future maintenance of these typical cold deck roofs.

First, for roofs situated over habitable rooms the felting should be in three layers laid in hot bitumen with the surface covered in bitumen-bedded chippings for protection from the sun. If chippings are not used the surface felt will lift and crack after only a short time, particularly on roofs with a south aspect. Periodic inspections should be carried out to confirm that the protective layer of chippings is in place and has not been washed away.

Secondly, the void underneath the decking must be provided with some ventilation. The decking itself may be of tongued-and-grooved or butt-jointed boards; it may comprise Stramit board, made of compressed straw, which was popular in the nineteen-fifties and nineteen-sixties; it may be of chipboard or some other type of manufactured boarding, either pre-felted, with only two additional layers of felt rather than three, or unfelted. The purpose of ventilation is to prevent an accumulation of trapped vapour under the decking which can otherwise cause lifting or 'blowing' of the felt. This may then indicate trapped water vapour in the void due to lack of ventilation. Ventilation may be provided quite easily by allowing for spacers under the perimeter fascia boards of small roofs and by setting vents into the surface at intervals where larger areas of flat roof are involved. Adequate ventilation is essential to prevent interstitial condensation.

Thirdly, the roof void should be protected from the entry of warm, damp air rising from rooms beneath (especially bathrooms and kitchens) by the use of suitable vapour barriers or vapour checks. This means that the ceilings beneath should be virtually sealed and airtight. If narrow ceiling cracks, holes or joints exist, particularly where electric wiring passes through, then warm, damp air will enter the void and condense on the underside of the cold decking in the roof void, a feature encountered mainly in wintertime. Polythene sheets may be used as vapour barriers provided that they are carefully lapped and sealed at all joints and are placed on the warm side of any insulation — generally this will mean directly over the ceiling plasterboard. Electric wiring should be run not through the sheets but under them so as to maintain the integrity of the vapour check. Foil-backed plasterboard is often used but in this case joints should be sealed. Metal foil, however, makes the best vapour barrier.

Fourthly, the roof should be insulated. Experience indicates that flat roofs tend to suffer greater heat losses in winter and heat gains in summer. This may be reduced by the provision of ceiling insulation which should be laid over the vapour barrier or vapour check but with an adequate ventilation gap to permit free circulation of air directly under the decking.

Fifthly, special care is needed to maintain the flat roof detail in good order since most problems start with small leaks at flashings, upstands and parapets and at points where openings are formed for soilpipes, flues and the like. Warm boiler flues will present a special problem since the built-up felt around them will be softened by the heat and the weather-sealing will often be deficient.

Where the roof forms a dished tray with an internal rainwater pipe, frequent inspections would be advisable to clear leaves and debris, especially if trees are near, since if the internal rainwater pipe blocks, the tray will fill with water which could then penetrate the decking at any weak points around the perimeter.

Well-constructed built-up felt roofs have a useful life span of 10 to 15 years, after which renewal of the covering will generally become essential. Occasionally a felt roof may last longer, especially if sheltered from the sun, but poorly-constructed felt roofs often have a shorter life.

A problem arises in giving advice on the action to be taken where a flat felt roof is weathertight but is of an age when re-covering may be anticipated. To some extent local repairs in the form of patching defective areas and renewing flashings and upstands are often possible and may prolong the overall life of the roof. The main problem arises from the fact that small leaks, often undetected for some time, will cause rapid deterioration of the decking and timbers beneath and destroy the effectiveness of insulation and the vapour barrier. In these circumstances the condition of decking and timbers cannot be accurately assessed without full exposure.

Stramit board decking will absorb moisture and convey it over a wide area by capillary action. If ventilation is poor, this warm, damp atmosphere will encourage wet- or dry-rot fungus to grow. Chipboard deckings will also absorb moisture and deteriorate rapidly, often without any moisture reaching the ceilings beneath to give warning to the occupants that all is not well above. Conventional butt-jointed or tongued-and-grooved boarding have, in the writers' experience, proved more durable. Any decay occasioned by small leaks in the surface covering has tended to be more localised than would be the case with chipboard or Stramit, although boarding is often more prone to outbreaks of wet- or dry-rot if it is unseasoned and ventilation is poor.

The re-covering of a flat felt roof in new built-up felt to British Standard 747 will be a fairly straightforward building operation and, for a small area, can be completed rapidly. If a small domestic roof is involved it is generally possible to strip off the old covering, have one or two layers of felt laid and the joints sealed in one day, thus avoiding the need for weather protection in the form of temporary canopies and scaffolding. However, if the roof area is large, or the decking and joists need repairs, then the roof will be exposed for some time and the cost will have to be increased to cover not only the need for weather protection but also the repairs to the joists and decking. In such circumstances, defective flat roofs can cause considerable inconvenience to a property owner and result in substantial expenditure to put things right.

The purpose, therefore, of periodic inspections of the roof surfaces and ancillary items is to pre-empt any trouble in the form of small leaks which could result in major expenditure later on.

Thatch

Occasionally a surveyor may be asked to report on a thatched roof. There are some thatched properties in suburban areas and a considerable number in rural locations, probably more than one would expect. Thatch gives considerable advantages as a roof covering. It has good insulation properties against both heat loss and noise, and the noise of rain on the surface goes unnoticed. Generally a deep eaves overhang is provided, obviating the need for gutter or rainwater pipes and rainwater is allowed to run off the eaves overhang well out from the main walls of the building.

Vermin infestation in the thatch can be a problem. Birds cause considerable damage during the nesting season and squirrels and mice are very destructive to thatch once they have gained an entry. Many thatched roofs are covered with a fine galvanised steel wire mesh to prevent birds or vermin removing or burrowing into the straw or reeds but this can make access very difficult in the case of a thatch fire and a PVC or nylon mesh is now often used, melting when heat is applied.

Clients buying a house with a thatched roof should be warned that insurance premiums may be higher in respect of fire cover on both building and contents. A specific point should be made of notifying the insurers when a roof is thatched, this being a matter relevant to the risk. Certain obvious precautions should be taken by the property owner. He should not light garden fires near the building or use blow lamps for stripping paintwork near the roof and, in an urban location, some check should be made on Guy Fawkes' night and similar occasions and a hosepipe kept handy in case of ignition by fireworks or bonfire sparks.

Further, warning should be given concerning the estimated life span of the particular material. In the case of reed thatch about 60 years is typical but straw thatch has a much shorter life. During the life of the thatch the covering should be periodically inspected by a thatcher and may need an overhaul, generally taking the form of combing the surfaces, making good defective areas and renewing hips, ridges and details where these have deteriorated. A property owner should cultivate an association with a local firm of thatchers or, if no such firm is available, contact the nearest available rural locality where thatching skills are still practised.

When thatch reaches the end of its useful life it will need replacing in its entirety. Replacement may be undertaken with a different covering such as slates or tiles or with new thatch if thatchers are available. In many cases re-thatching will be realistic and not more expensive than tiling, bearing in mind that tiling will certainly involve new battens and sarking felt, renewal of or addition to rafters and the provision of gutters, rainwater pipes, fascias, soffits and different detailing around stacks. If the property is in a conservation area, or is a listed building needing special consent from the local planning authority, it may be that a change in the type of roof covering will not be permitted.

The fire risk associated with thatch will obviously be greater in older timber-framed buildings where the ceilings and walls may be of lath and plaster or similar construction which is itself a potential fire hazard in some circumstances. The need to maintain the electrical equipment in a property with a thatched roof in good order, and to test the wiring regularly, is self-evident. This is extremely important since when the thatch is renewed or repaired, the old thatch is often left lying in the roof void. This dry, brittle, dusty material is highly combustible, particularly in the case of straw thatch.

One final warning concerning thatched roofs is the prevalence of insect infestation - in particular furniture beetle and the death watch beetle where oak timbers are involved. There is hardly a thatched roof in the West Country which does not contain some form of insect infestation, usually *anobium punctatum*.

Roof coverings in general

Lead is a traditional covering for both flat and pitched roofs with the life dependent upon the lead thickness adopted. Heavy cast lead was used in period houses, churches and other buildings and may last hundreds of years. Modern leadwork is generally the thinner milled lead which is more likely to suffer from 'creep' on slopes, flashings and other details, and from splits in exposed locations or where subject to movement. Generally a life span of 80 to 100 years is typical for modern leadwork. Modern lead is clearly preferable to mineral felt for flat areas and detailing, and the author prefers to specify leadwork where possible if the client's pocket is deep enough, especially in locations with a southerly aspect where mineral felt tends to rapid deterioration.

Large areas of leadwork without rolls or drips at suitable intervals may fail due to thermal movements which cause rippling and splitting. In its original state the lead is a bright silver colour but on contact with the atmosphere this changes to the familiar matt grey due to a coating of lead carbonate which forms on the surface. This surface carbonate protects the lead from further deterioration. The only chemical deterioration likely to occur is from strong acid attack and acid can occur in rainwater if this is discharging on to the lead from a roof area where there is moss growth, or material subject to moss such as ageing shingles. A point discharge such as regular dripping from a mossy roof will cause pitting to the leadwork due to such an attack.

Zinc was often used as an alternative to lead for flat roofs and detailing, especially in chimney stack flashings and soakers and small flat bay roofs in traditional dwelling construction. In pollution-free localities the life of such zincwork is generally in the range of 20 to 40 years and sometimes even longer. Good detailing to zincwork is essential for durability, and the rolls and drips are necessary to accommodate thermal movements. When chemical pollutants are present in the atmosphere, the life of zinc is shorter and failure in the form of pitting and splits is possible after as little as 15 years. As with leadwork, a particular problem can arise with a constant drip of rainwater at a specific point over a period causing the acid in the rainwater to wear away the zinc and create a point for leakage. Zinc, like lead, also develops a carbonate on the surface which affords some measure of protection but this is thinner than the lead carbonate and does not provide as effective protection, especially where the atmosphere is polluted by sulphur dioxide.

Asphalt is often used for better quality flat roofs and is more durable than built-up felting provided that the surface decking is sound and free from excessive thermal movement. A life span of 20 to 40 years is generally found to be possible, depending upon the quality of the original job and the degree of exposure to sunlight. Deterioration is indicated by the appearance of surface splits and cracking and by surface bubbles which form from the expansion of vapour trapped in the decking. Minor surface crazing may not of itself indicate general deterioration since the surface asphalt is trowelled into place and the trowelling is often overworked, as it is on stucco renderings which craze for similar reasons. This overtrowelling brings to the surface a skin of bitumen which may tend to craze due to slight differences in rates of expansion and contraction on the surface compared with the deeper material.

A normal thickness of asphalt is about 20 mm and anything less than this may be regarded as sub-standard and of a potentially shorter life span. Asphalt is not a suitable material for point-loads and will indent and spread under the weight of water tanks and other imposed weights. Very often considerable damage can be done to an asphalt roof by indentations created by chair legs, flower tubs, duckboards and the like. Most problems arise with the asphalt detailing. If the material is laid at angles to slopes and vertical upstands without suitable tilting fillets, creep will occur, especially in a south-facing location.

Copper is a traditional roof covering used for small areas of flat roofs for detailing, and for pitched construction in domes and spires where the distinctive green patina is required for aesthetic reasons. Copper sheet used for this purpose is thin but surprisingly durable, although not quite so durable as lead. In time the metal ages from the effects of thermal movements and splits and ripples to the surface which will appear indicate that renewal is required. Rainwater with a high acid content attacks copper in the same way as with zinc and to a greater degree than in the case of lead. The thin metal - especially the rolls - is prone to damage from any foot traffic or ladders which may be used for maintenance.

Detailing

As a general rule all flat roofs should have a natural drainage fall and should not hold water. The reason for this is that ponding on the surface, especially with asphalt and built-up felt, will result in marked differences in temperature between the wet cold areas and the dry warm areas, especially in warm sunlight following overnight rain. This causes relative thermal movements between cold and warm parts which result in rapid deterioration of the surfaces. Generally, in new construction a fall of at least 1-in-40 is desirable to allow for possible bedding down of the construction in later life and to accommodate any minor deflection in timber joists. In existing flat roofs where any bedding down and deflection has ceased, a shallower fall may be acceptable but this should not be so shallow that the surface holds water.

Experience tends to indicate that most problems with roof construction arise at the detailing rather than the main areas of covering. Rain penetration of flashings and soakers of stacks and fire walls or to upstands and flashings at the perimeter of flat roofs is common. On close examination many roofs will exhibit a history of repairs to such details, the repairs often being of a highly unsatisfactory nature. Unfortunately the average building owner or lessee will rarely carry out any close inspections himself and in the event of a leak will rely upon the judgment of a roofing contractor. The eventual bill will be paid without knowing whether or not the work done was necessary, suitable or reasonably priced. Often it is only when the building is subjected to the scrutiny of a surveyor that the nature of the past repairs comes to light.

Chimney stacks and copings

Particular problems may arise with downward damp penetration through soft brickwork and mortar in chimney stacks and parapets. The design of stacks and parapets is often poor and for the correct construction of parapet wall copings two conflicting requirements have to be met. First, copings should be suitably mounted on impervious damp-proof courses and should be adequately weathered with overhangs and throating. Secondly, copings should be firmly bonded in place with mortar joints which will not deteriorate and be built of materials which are sufficiently durable and resistant to frost damage. Since the damp-proof course beneath the coping is a point where adhesion is poor, the design of copings must be a compromise between these two requirements. All brickwork to copings and

Figure 6.2 Clients will need to know about loose pots, porous pointing and leaking flashings. A good push or a strong wind are all this stack needs to cause collapse

stacks above damp-proof courses where exposure is continuous should be in special rather than ordinary quality brickwork since the latter will suffer frost damage in time. Similarly, tiles, coping stones and other finishes to copings must be of a suitably durable type for the conditions of exposure involved.

When chimney stacks are redundant and clients are quite sure that they will not be required again, an opportunity arises to remove the stack to below the roof line and to tile or slate over. This is often the most suitable solution as an alternative to rebuilding where a stack is in poor condition. When stacks are disused but in adequate condition they may be retained provided that suitable ventilation is applied at the top and bottom of the flues. If an unventilated cap is used or no ventilation is provided at the base of a flue, weather penetration downwards coupled with the effects of condensation within will often result in deterioration to the parging and brickwork and brown damp stains appearing on chimney breasts inside.

Gutters

Generally, all good roof design should incorporate the provision of adequate guttering and rainwater pipes to carry all roof water away to a storm drainage system. It is important that stormwater be taken away from the perimeter of the building to prevent damp penetration through walls and to avoid subsoil erosion from continual discharge of roof water into soil adjoining the main walls.

Certain traditional roofs encountered in period construction will not have gutters or downpipes, particularly in the cases of thatch and traditional stone slab construction where a deep eaves overhang is provided instead. It is vital with this type of construction that the area around the building should be paved, with a suitable drainage fall away from the main walls and with a good clearance between exterior and interior levels. A rendered plinth treated periodically with a bituminous water-proofing paint would also be advised. An examination of a number of thatched cottages over the years will reveal that dampness at the base of the main walls, especially on the weather-sides, will invariably be a problem unless the run-off of water from the roof slopes is adequately catered for. A combination of thatched roof, solid rendered walls of stone and lime mortar, and high exterior ground levels is often encountered; not surprisingly in such circumstances, invariably there will be a damp problem at the base of walls internally with a history of plaster repairs, or false walls applied over damp plaster.

A client proposing to buy a cottage with a thatched roof may have seen it through rose-coloured spectacles, especially if he first viewed in summertime. Such a client may have thought it quaint that visitors step down into the building instead of up, and will view uncritically the moss-covered and loosening rendering around the base of the building. The vendor's panelling to dado height around the ground floor rooms could be considered part of the charm. It may not be until the following winter that the client begins to detect the effects of damp penetration at the base of the walls internally and that the carpet underlay is becoming mouldy around the skirtings. Perhaps then the significance of the high exterior levels and the nature of the construction will be appreciated and the client may feel inclined to look out the surveyor's report to see what was said on these subjects.

Often, small projections devoid of gutters or rainwater pipes are built and this will be commonly encountered in small bay roofs and porches of all ages. No matter how small the projection, guttering and downpipes are always advisable unless the roof overhang is very deep. Where these are omitted it will often be found that exposed sills and woodwork beneath have deteriorated from the effects of dripping rainwater. Staining down walls is also commonly found where guttering is omitted and this can cause considerable discolouration on north sides where moss growth is then encouraged.

Rainwater will occasionally be found taken to water butts instead of to stormwater gullies and this may be unsatisfactory, especially if the roof area is large, since overflowing from the butt will be undesirable. If water butts are needed it is suggested that gullies also be provided to take surplus water from the butts.

Modern uPVC rainwater systems have been found durable and maintenance-free if correctly installed with adequate brackets and sponge seals. Older rainwater systems will be encountered which may be less durable or suffer other drawbacks. Cast iron can be long-lasting if properly maintained but maintenance is often expensive and consequently poor and the hidden surfaces such as gutter interiors are often left unpainted. Asbestos-cement rainwater systems appear to have a useful life span of about 30 years and then become porous. Pre-cast concrete 'Finlock' gutters, commonly used in the nineteen-fifties, need periodic internal bituminising to avoid downward penetration of water into the walls beneath, especially at joints. Where an existing rainwater system needs repair and repainting it is often advisable to consider complete renewal in uPVC since the cost is often comparable and the final result should require no further maintenance in the foreseeable future.

Chapter 7

Joinery and carpentry

The woodworm is some of the worst I have seen. The only thing that is holding the roof together now is the fact that the grubs are holding hands in the holes
Surveyor to worried client

The survey of accessible joinery and woodwork is intended to reveal possible timber defects and to make recommendations accordingly. The surveyor should be conversant with the main species of wood-boring insects and wood-rotting fungi although he will undoubtedly encounter unfamiliar wood-borers from time to time and may often have difficulty in identifying the fungus responsible for timber decay in the early stages when no fruiting growths have yet appeared.

Wood-boring insects

The common furniture beetle (*anobium punctatum*) is probably present in about 80% of pre-War houses in the London area and no doubt in a higher proportion of pre-1914 buildings. It is now becoming evident even in timbers in the earlier post-War houses dating from the nineteen-fifties.

Evidence of activity will be small circular flight holes, smooth inside and accompanied by deposits of bore dust – generally to be found in softwood roof timbers, floor timbers and staircases and especially prevalent in plywood panelling under stairs and to joists and sleeper plates in suspended timber ground floors of inter-War origin.

The furniture beetle normally attacks only sound wood, although it is encouraged by timber which has been affected by damp. The flight holes are generally about 1.5 mm in diameter and the bore dust contains small faecal pellets which give it its characteristic gritty feel.

The death watch beetle (*xestobium rufovillosum*) is often present in old hardwood beams and other hardwood timbers in period structures. On occasion, fungal attack may be required for this insect to become active so that damp conditions and poor ventilation would normally be an initiate for attack. Mostly oak is affected and many oak beams used in period buildings were formerly used elsewhere, either in other buildings or as ships' timbers, so that they were already second-hand, perhaps carrying with them death watch beetle grubs when first used.

Figure 7.1 Common furniture beetle

Figure 7.2 Undersides of Victorian floor boards showing flight holes of the common furniture beetle. Interestingly, no flight holes were apparent in the top surface of this board. (Diameter of coin 23 mm)

Death watch beetle flight holes are larger than those of the common furniture beetle, being 2 to 3 mm in diameter. The bore dust contains characteristic ovoid pellets.

Wood-boring weevils (*pentarthrum huttoni* and *euophryum confine*) infest damp and decayed wood where fungal attack is also present. The flight holes are smaller than those of the common furniture beetle, being less than 1 mm in diameter. Weevil attack will commonly be found in the ends of joists and sleeper plates under suspended timber floors, and in other situations where there is contact between the

Figure 7.3 Death watch beetle

Figure 7.4 Death watch beetle damage

timbers and damp masonry. Weevil attack often causes the complete distintegra-tion of a damp sleeper plate or joist end resulting in the partial, if not total, removal of support from a floor.

Bark borer (*ernobius mollis*) attacks the bark and sapwood of softwood timbers and is often found in roof rafters and joists. The flight holes and bore dust are similar to those of the common furniture beetle but if bark borer is confirmed, the

Figure 7.5 Wood-boring weevil

Figure 7.6 Powder post beetle

surveyor will be able to reassure his client that no specific remedial action is required since this insect is not structurally destructive and will eventually depart of its own accord from the areas of infestation.

The powder post beetle (*lyctus* family) attacks hardwoods, mainly oak, ash and elm, and especially elm. The insects are active only in the sapwood with the grubs reducing it to a fine powder. The activity of the powder post beetle may be distinguished from that of the common furniture beetle by the very fine, as compared with gritty, bore dust and the fact that activity is confined to the sapwood edges of roof timbers, boarding and other members. The elimination of powder post beetle requires the careful selection of timbers to exclude those with excessive sapwood before use.

The house longhorn beetle (*hylotrupes bajulus*) attacks softwood timbers in the timber yard and in buildings. Because it causes rapid deterioration in the affected wood this pest is notifiable,in certain areas. The flight holes are oval-shaped, 3 to 9 mm in diameter. A feature of the damage caused is that the timbers are eaten away internally, often with little external evidence to show the severity of the attack. At present this insect is generally found to be confined to buildings in the Home Counties but there are reports of it appearing further afield.

Figure 7.7 House longhorn beetle

For information regarding damage caused, distribution within the UK, and the area within which preservative treatment of softwood roofing timbers is mandatory for house longhorn beetle, see 'House Longhorn Beetle Survey', BRE Information Paper IP 12/82, July 1982.

The larvae of the digger wasp has been known to attack teak timbers, normally thought to be immune from such infestation, resulting in almost complete disintegration.

Various other wood-boring insects will leave flight holes in new wood, especially imported wood, and may be found in new buildings where such wood is used.

Pinhole borers may be occasionally found to have been active. However, some flight holes in new wood are permitted by current Codes of Practice since it would be unrealistic to discard useful timber merely because of a small number of old flight holes. In such circumstances it is sufficient to confirm that the timbers have in fact been suitably pre-treated before use and that all insect activity has ceased. CP 112, Part 2:1971, Appendix A.12 states that 'Scattered pinholes and small occasional wormholes are permissible in all grades. All pieces, however, showing active infestation should be rejected or subjected to preservative treatment in accordance with CP98, 'Preservative Treatment for Constructional Timber', (superseded by BS 5268, Part 5:1977).

Wood-rotting fungi

Wet-rot is normally associated with the fungus *coniophora puteana* but there are several other fungi capable of causing the symptoms of wet-rot and identification of the actual fungus responsible will be difficult until an attack is at a severe and late stage.

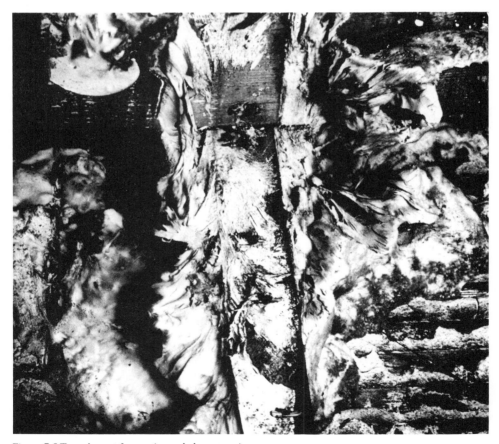

Figure 7.8 True dry-rot fungus (*serpula lacrymans*)

Figure 7.9 Wet-rot (*trametes serialis*). Brown fungus with cuboidal cracking

The main factor distinguishing a wet-rot attack from the true dry-rot (*serpula lacrymans*) is that wet-rot requires damp timber in order to become active and spread and will confine itself to the affected and immediately adjacent timber. Dry-rot requires a humid atmosphere to become established, and damp timbers in situations lacking ventilation are ideal targets for attack. Once established, however, dry-rot fungus will extend readily into adjoining sound timbers and will travel through masonry and plasterwork in order to reach other sound timber which then in turn becomes infected.

Wet-rot is most often found in external joinery, especially lower window frames and sills, door posts and wood thresholds in contact with the ground. Garage door posts and fascias subject to driving rain penetration can also be affected. It is also found in roof timbers around stacks where weather has penetrated and in cellars and sub-floor areas subject to dampness. In an advanced stage the fungus will produce dark brown or black strands similar to old cobwebs and these will often be found forming patterns on the wall plaster in damp cellars.

Wet-rot is very common in the painted softwood window units used in the nineteen-fifties and nineteen-sixties due to the lack of adequate treatment of such timber before use. Often complete replacement of such window joinery will be found to be necessary due to this problem. In coastal areas and areas of high exposure this defect is quite general. In certain districts reference may be made to 'south coast rot' which is essentially due to difficulties arising from exterior woodwork of poor quality used in post-War speculative developments. One avenue of investigation open to the surveyor during his inspection is to probe carefully the exterior woodwork, paying particular attention to the obvious areas at risk and including especially any porches, conservatories or similar timber-framed structures, particularly those facing the prevailing weather.

Figure 7.10 Constant saturation by water from defective roof overflow has caused serious wet-rot in this painted softwood bay window. The south-west facing location is also subject to the prevailing weather

Remedial measures in the event of a wet-rot attack in exterior woodwork should be made by cutting out affected wood well back from the point of decay and splicing in new pre-treated timber, preferably also using pre-treated wooden dowels and connecting with suitable tenons. Considerable care is required to achieve a proper repair and if wet-rot is extensive, this is often not justified. When rot is widespread, complete replacement with new units will generally be preferred. Filling repairs to exterior woodwork affected by rot will often be encountered in practice and such repairs must always be regarded as unsuitable as the decay will invariably return. Filling repairs, painted over, could be used by a vendor to disguise the nature of the rot and with a view to helping the sale of the property.

When the property under review is suffering from attack by the true dry-rot fungus, the necessary remedial action will include not only the replacement of all affected wood with new pre-treated timber but also a suitable programme of treatment for adjoining timbers, brickwork, plasterwork and sub-floor areas well away from the actual point of decay. Considerable care is required in devising a suitable specification for such works to ensure that the fungus is eradicated.

Generally a dry-rot fungus will be more likely to attack sub-floor timbers, spread unnoticed behind panelling and inside flat roof voids, all being points where a warm damp atmosphere may arise due to poor construction and lack of ventilation. Such conditions will mostly be found in older buildings where the amount of timber built into the structure may be considerable. Timbers such as those in stud partitions, lintels and wall plates may provide a passage through the structure which the fungal mycelium can travel, sometimes for considerable distances. The dry-rot fungus is able to spread in this way and attack dry timber because of its ability to produce water-carrying strands, or rhizomorphs, which are modified hyphae and are vein-like structures which can be as large as 10 mm in diameter. These rhizomorphs convey water from damp wood into dryer wood elsewhere, passing over masonry and through mortar joints in walls. They also 'store' food so that even if the rotted wood is removed the rhizomorph is still able to grow and infect new wood.

The mycelium of the dry-rot fungus is a snowy white colour (as compared with the brown or black strands of the wet-rot) and grows rapidly in damp humid conditions. Where the edge of the mycelium is in contact with light it turns bright yellow. Timber decayed by the mycelium exhibits deep fissures and the wood breaks up into cubes with the cuboidal cracking becoming apparent on the surface. Eventually when an attack is well established the sporophore or fruiting bodies will appear, these being thin and pancake-like, white around the edges, giving off rust-coloured spores and a strong mushroomy odour.

The sporophore or fruiting bodies of the wet-rot fungus are rarely encountered indoors but may occasionally be found in decayed external fascias and sills. They are smaller than those of the dry-rot fungus, generally brown in colour and in the form of thin plates attached to the face of affected wood.

In modern structures, one particular area where timber decay from either wet-rot or dry-rot is possible is within the void to a flat roof, especially if this is of sub-standard construction. Built-up felt roofs will often be found without any void ventilation and lacking any effective vapour checks. The climate within such roofs is ideal for the rapid growth of these fungi. The decking material may be Stramit board or chipboard which conveys water from a small surface leak by absorption through the roof area without any staining becoming apparent to the ceilings beneath. In such circumstances the dampness in the roof void, coupled with the warmth from the sun on the roof surface, will produce very rapid deterioration so that eventually when the roof covering has to be stripped for renewal, renewal of the decking and joists may also be found to be required.

An inspection of timber members used within the structure of a building can rarely be complete even if the building is empty of all contents and floor coverings and exposure of floor joists in all areas is possible. The reason for this is that some timber surfaces may be permanently hidden and lie within the structure in such a way that precludes a full inspection of all surfaces without taking the structure apart.

Timber-frame construction

A surveyor may be asked to prepare a report on a timber-framed building, that is a building in which the outer walls are constructed in load-bearing timber-work instead of traditional brickwork, blockwork or stone. Construction time can be much shorter on site since the panels will generally be delivered in a partially assembled form. Also the standard of thermal insulation can be higher and the cost of the main structure lower than would be the case with traditional construction.

In 1996 the Building Research Establishment published three reports confirming the long-term durability of timber frame dwellings built between 1920 and 1975 (Timber frame housing 1920–1975: inspection and assessment. Timber frame housing systems built in the U.K. 1920–1965. Timber frame housing systems built in the U.K. 1966–1975).

In these reports the BRE found that the incidence of timber decay in the dwellings inspected was generally very low and with a few exceptions any decay was localised, superficial and usually the result of poor workmanship or inadequate maintenance. The BRE concluded that the performance of timber frame dwellings built between 1920 and 1975 is generally similar to that of traditionally built dwellings of the same age. The two reports on systems provide detailed guidance on the identification, constuction and performance of 54 individual timber frame housing systems. The BRE concluded that, provided that regular maintenance is carried out and repair and rehabilitation work meets accepted levels of good practice in design and construction, a performance comparable with traditionally constructed dwellings of the same period should be maintained into the foreseeable future.

The survey of an existing timber-framed building presents certain difficulties as all important structural elements are hidden. There are nine main considerations in good timber-frame design which are:

(1) The design of the main structural members and fixings to resist all forces to which the building will be subject;
(2) The adequate pre-treatment of timbers against rot and beetle attack;
(3) The provision of damp-proof courses and membranes;
(4) The provision of vapour checks in the wall panels to prevent interstitial condensation;
(5) The provision of barriers against weather penetration;
(6) The provision of adequate external claddings;
(7) The provision of adequate thermal insulation within the panels.
(8) Breather membranes in the walls to allow trapped moisture to escape.
(9) Fire checks in the cavities of connected timber frame buildings to prevent the rapid spread of fire within the structure.

None of the foregoing matters can be checked from a superficial examination of the wall surfaces except item 6. A surveyor inspecting a timber framed building may therefore have to discuss with his client the need for a structural check on the design from details available from the manufacturer and the possibility of opening-up selected points or using an endoscope to confirm hidden details. The local authority or other checking authority responsible for building control will normally insist that full structural calculations are submitted with deposited plans and that items (1) to (9) above are provided for before approval for a timber framed building is given; details deposited with the local authority may be available for inspection but copies cannot normally be taken without the copyright holder's consent.

Suspected variations in the actual building as compared with approved plans should be checked, especially in the sizes and locations of openings in main walls or variations in the layout of the timber panels. This is most important as the structural principles involved in timber-framed construction are quite different from those of traditional in brick, block or masonry building. Panels in timber-framed work require lining and diagonal bracing to give lateral restraint, unlike traditional work which obtains lateral restraint from the cross-bonding and intersection of walls. If lateral restraint in timber-framed construction is inadequate then the structure may be distorted parallelogrammatically by a process known as 'racking'. The danger of racking damage to a timber-framed building is at its greatest where openings are formed in the walls near to corners or if openings larger than specified are formed, thus reducing the overall stiffness of the framework. Detached timber frame buildings in areas subject to high wind loadings are most at risk.

The surveyor could ask to see structural design details and calculations for any modern timber-framed building which he is asked to survey. Clearly it would be advisable for the owners of such buildings to keep a copy of the details and calculations to be passed on to subsequent owners since questions are bound to arise from time to time when such properties change hands. In cases of doubt it is recommended that the details and calculations be referred to a consulting engineer for checking, and in any event a careful comparison should be made between the design and the actual building with a view to detecting any changes which may be structurally significant.

Structural alterations or conversions made to an existing timber-framed building should be undertaken only with the benefit of expert professional advice and if such a structure is found to have been modified subsequently, the nature of any modifications should be carefully checked.

The external brick or block skin used as a cladding to modern timber-frame designs is not of itself structurally important. It often comes as a surprise to clients to learn that a building which appears superficially to be built of brickwork is in fact of timber construction. In some designs the base of the brickwork is used to hold down the base of the adjoining panels using angle clips but brickwork or blockwork claddings are otherwise not normally structurally important to the design.

Preservation

It will be appreciated from the previous descriptions of the main wood-boring insects and wood-rotting fungi that good timber preservation in the form of pre-treatment is necessary to ensure durability in new work. A surveyor's knowledge of preservation techniques is important to assess the defects which will be found in older buildings and the remedies which may be proposed.

Different species of wood show great variation in their resistance to decay and beetle attack. Generally softwoods are most vulnerable; for example, ash and beech decay rapidly if conditions are suitable. Hardwoods are more durable in most cases and may be resistant to fungal attack even in conditions where fungal activity would be likely. Oaks and teaks, for example, are remarkably durable in damp conditions. In all timbers the heartwood is most resistant to decay and the sapwood least resistant. However, the sapwood will more easily absorb preservatives so that preserved sapwood of a non-durable species can be more durable than unpreserved heartwood.

Preservation is partly a question of good design to ensure that all timber members are kept dry and are not in contact with damp oversites or masonry. The climate around timber members should be cool, well ventilated and free from excessive water vapour. Areas most at risk will be suspended timber ground floors and enclosed roof voids, especially under flat roofs. In timber-framed construction the wall panels also require careful design, especially to give protection against weather penetration, provide vapour checks and allow entrapped moisture to escape through breather membranes.

Preservation also covers the treatments and surface coatings which may be applied to timber. Any water-proofing surface applied to timber will assist in preventing fungal decay since this will present a barrier to the absorption of moisture. Exterior woodwork is normally then painted in an oil-based paint, generally with two coats plus primer. Well-painted exterior surfaces are protected from moisture absorption by the paint and this degree of protection will be good if joints and putties are carefully sealed and painted over in such a way that no small passages for moisture remain. Painted surfaces reflect sunlight but surfaces with a hard clear finish such as varnish absorb the sunlight to some extent and this generally causes the surface to crack and peel quickly, especially on south-facing elevations.

An alternative to painting or varnishing is to give the exterior woodwork a periodic preservative stain finish so that instead of covering the wood with a surface weather-resistant shell, the treatment consists of regular brushing or spraying with absorbent preservative which soaks into the wood. It is possible that exterior woodwork treated periodically with an absorbent preservative stain will become more common in the future as the cost of traditional painting continues to rise.

To prevent fungicidal activity a wood preservative must so permeate the wood that fungal hyphae are unable to feed on it. To prevent insect activity within the timbers a wood preservative requires insecticidal defence properties against that specific species for which protection is required. Different wood-boring insects have different susceptibilities. Different treatments may be required for timbers where insects are known to be active and timbers where pre-treatment against future activity is required. In the former case eggs, grubs, pupae and adult insects may all be present in the wood and require eradication; in the latter case the protection required is protection against the ingress of egg-laying adults.

Chemical treatments should take a permanent or semi-permanent form to protect the timbers from fungal and insect attack for a suitably long period. Generally 30 years' protection would be a good standard at which to aim. This is possible because the chemicals are absorbed by and retained within the wood to a considerable depth and for long periods, unaffected by sunlight or other causes of deterioration.

Timber preservation may be undertaken in any one of five ways using chemical preservatives, these being: (a) Pressure-impregnation, (b) dipping, (c) steeping, (d) spraying, (e) brushing.

Pressure-impregnation involves the use of sealed tanks of preservative into which timbers are placed, the wood preservative then being forced into the timber under pressure. This is suitable only for pre-treatment.

Dipping involves immersing timbers in an open tank of preservative for short periods, perhaps a few minutes, and then allowing the excess preservative to drip from the wood.

Steeping involves treatment similar to dipping but extended over period of hours, days or even weeks, depending upon the nature of the timber involved and the degree of penetration required. Dipping and steeping are suitable only as pre-treatment before timbers are used.

Spraying may be used as treatment for *in situ* timbers. Generally a coarse spray nozzle is used so that droplet size is large and rapid saturation of the wood surface is obtained without producing a fine airborne spray which could prove unpleasant or hazardous to operatives.

Brush treatments are often less effective than spray treatments since areas which are inaccessible to a brush may be reached using a spray nozzle, especially in suspended floors, behind panelling and in the extremities of roof voids.

Chemical wood preservatives are classified according to three main groups (and by reference to British Standard 1282:1959 (revised 1975)) as being tar-oil types, organic solvent types and water-borne types, and these are further sub-divided in accordance with the main chemical base of the preservative.

Treatment

Faced with an existing building which suffers from the effects of dry-rot, wet-rot or attack by wood-boring insects, a surveyor will have to make some recommendations as to a suitable course of treatment in addition to commenting upon mere facts and describing their significance.

It is suggested that the existence of wood-boring insects in timbers where the timber strength is not affected and where the activity is indicated by scattered flight holes only is best treated by spraying timbers until rejection, with an insecticidal preservative of a type appropriate for the species of wood-borer involved. The question of how far such treatment should extend into areas of adjoining timber construction is difficult, but in residential property, for example, it may be most sensible to confine treatment to the immediate area of infestation but at the same time to advise the client to have the other areas checked again after five years have elapsed. Thus it may be possible to deal with the fact that flight holes indicate that insects have emerged to lay eggs elsewhere, the results of which could take some years to appear in the form of new flight holes. If timbers have been structurally weakened by wood borers it will, in addition, be necessary to recommend replacement or strengthening of weakened timbers, paying particular regard to the ends of floor joists and the stiffness of principal members in roofs.

A suitable specification for wet-rot attack may be prepared on the assumption that if the cause of damp can be removed and the timber allowed to dry out, no further attack will develop in timbers not already attacked. The specification should then include the renewal of all affected wood well back from the point of decay, the protection of surrounding timbers (necessary to prevent decay arising in the future) and the eradication of the source of dampness where this has been traced and found to have caused the attack. All new wood should be spliced and jointed into place to provide a lasting repair, free from weak points externally or inadequate weather protection. All replacement wood should be adequately pre-treated.

Specifications for dry-rot control require rather more details if these are to prove suitable, having regard to the particular nature of this fungus as already discussed.

Depending upon the circumstances, a suitable specification should generally include the following specific provisions:

(1) Cut out, carefully remove from site by the shortest practicable route and burn all the following: timbers showing cuboidal cracking, brown colouration, white mycelium or soft areas when probed; all sound timbers within 1 m from any of the foregoing defective timbers; all debris within roof voids, sub-floor voids or other areas adjoining, particularly timber off-cuts and similar debris.
(2) Remove wall plaster, rendering, skirtings, wall panels, ceilings and other coverings as necessary to trace full extent of fungal activity in adjacent construction. Carefully clean out all exposed areas and clean down all surfaces within a distance of 2 m from actual fungal activity. Sweep out all sub-floor voids, roof voids and similar areas. Remove all dust and debris from site by the shortest practicable route and burn.
(3) Blow-lamp all masonry surfaces within an area of 2 m from area of fungal activity until surfaces are too hot to touch. When cooled, liberally apply fungicidal solution to all such surfaces.
(4) Apply fungicidal solution to all remaining woodwork within 2 m from point of fungal activity in two liberal coats.
(5) Make good to all timberwork in pre-treated wood which has been first given two good coats of fungicidal solution on all surfaces.
(6) Make good disturbed wall and ceiling surfaces using 5 mm of floating coat of zinc oxychloride plaster over the rendering coat in the area of fungal attack and for a distance of 300 mm surrounding such area. Apply zinc oxychloride paint to painted surfaces where plaster is not to be applied and over a similar area and distance from previous attack.
(7) Make good all design defects responsible for original dry-rot attack, including additional airbricks and improved ventilation, renewing defective damp-proof courses, preventing moisture penetration, eliminating condensation, clearing the bridging of damp-proof courses, repairing roof and plumbing leaks and timbers in contact with damp oversites or masonry as necessary and in accordance with good building practice.

During a survey and when advising on timber treatment it is often helpful to be able to confirm the location and spacing of timbers behind surfaces in stud or similar construction. In addition to tapping the wall to check for hollow areas a useful tip is to use a compass which will indicate the location of steel nail heads and may enable the layout of timbers to be confirmed without exposure. Alternatively, metal detectors and stud sensors are now inexpensive and could form part of the survey equipment.

Finally I would emphasise the importance of exercising prudence when advising clients to invest in chemical treatments. Some of the chemicals used now and in the past may be found to have damaging effects on the health of the operatives using them and the subsequent occupiers of the buildings treated. There is growing evidence to suggest that a great deal of unnecessary timber treatment may be carried out, much of it on the advice of surveyors and valuers who may find isolated evidence of furniture beetle or other infestations and refer clients to specialist contractors who will undertake extensive spray treatment throughout the building using Lindane, Pentachlorophenol and other chemicals. We may expect to hear more about some of these chemicals in the years ahead as their long-term effects become better understood (see Page 251).

Chapter 8

Finishes and surfaces

All of a sudden, there was a noise like an express train running through the house. The wife opened the door to the hall and found that the whole of the landing ceiling had come down. The mess was unbelievable. The last time we saw anything like it was during the Blitz.
Tenant's letter to landlord

Plaster finishes

In most parts of the UK, dwellings built after 1938 will have ceilings of gypsum plasterboard. Exceptions to this will be found occasionally where timber laths and plaster or steel-mesh lathing and plaster have been used, but plasterboard will be the more normal. Prior to 1938, timber laths and plaster were generally used.

The life of the older lath-and-plaster ceilings will depend upon the quality of the original work and the degree of exposure to dampness from roof and plumbing leaks. Lath-and-plaster affected by serious leakage from any source will invariably have been weakened and will require renewal. During the course of the survey the condition of any lath-and-plaster ceilings should be checked by probing from below and, where possible, examination of the plaster key from above. If loose, soft or 'off-key' areas of ceiling plaster are found, general renewal in plasterboard should be recommended. Most lath-and-plaster ceilings will now have a limited remaining useful life and in most cases clients should be encouraged to budget for renewal in plasterboard.

Occasionally, fibreboard ceilings will be encountered, particularly in older buildings which may have been war-damaged during the Second World War, and also in some nineteen-thirties houses. This is inferior to plasterboard. It is not fire-resistant and is soft and easily damaged. Any chemical spraying of roof timbers for furniture beetle disinfestation undertaken over fibreboard ceilings will result in stains appearing beneath due to high absorbency, a point which should be borne in mind with this material.

Suspended ceilings of various types are commonly employed in commercial buildings and will occasionally be encountered in dwellinghouses. In most cases there is no problem about lifting a panel or two to examine the void behind and this can be quite instructive as that void will probably be in its original undecorated condition with a past history of defects on view and services exposed.

Wall plasters in modern buildings should be checked for loose or shelling areas by tapping and probing as necessary. Problems of shelling wall plaster can often arise in speculative construction if walls are plastered when they contain a high water content and if the plaster used is almost impervious when set. Occasionally major or complete replastering of new houses becomes necessary from this cause. The shelling and bulging of plaster finishing coats on clinker or breeze block walls and partitions generally arises about six to nine months after completion. Replastering of the defective finishing coat can be difficult because of the high suction of the undercoat. The Building Research Establishment published a note on this topic in 1971, 'Shelling of Plaster Finishing Coats', BRE Information Sheet TIL 14 1971.

Wall plasters which are well set after the first year or so in the life of the building should prove durable, but plasterwork in older buildings will become defective in the presence of dampness. Soft, damp and defective wall plaster in older buildings should be reported upon with the necessary advice. Where replastering is required, any special requirements should be mentioned, including the need for special plasters to be used where past dampness could give rise to efflorescence and surface staining of the new plaster.

Other wall and ceiling finishes

Walls and ceilings can have a wide variety of finishes other than plaster skim. Ceilings may be of plasterboard with scrimmed joints and may have a textured coat of Artex or similar finish. This will absorb the slight movements which concentrate in plasterboard ceilings and would otherwise cause a pattern of hairline cracks to the joint lines; it will not absorb more severe shrinkage cracks, however, and anything greater than hairline will probably appear eventually through the textured coating. Various patterns of swirls, combing and other effects are possible with textured coatings, depending upon the skill of the plasterer.

Walls can also be given a textured coating over existing plaster skim. In older construction, a stippled wall plaster may occasionally be encountered as an original feature. Whereas the modern textured coatings are slightly flexible, the older type of stippled plaster finishes do not have any flexibility and tend to be rather brittle. As a consequence, these stippled plasters to walls and ceilings may suffer cracking and the cracks may be difficult to eradicate or disguise since a stippled finish cannot be papered over with lining. If a smooth backing is required for wallpaper, the stippled surfaces have to be removed and the walls or ceilings re-skimmed in a new surface coat which could prove a major exercise in a house where all walls and ceilings are so treated.

Generally, plasterboard ceilings which are given scrimmed joints and a thin board-plaster finish will tend to suffer from a pattern of minor shrinkage cracks following the joint lines in the plasterboard panels. This arises more in ceilings to top floor rooms under timber joists due to the tendency for lightweight ceiling joists in roofs to be used for occasional maintenance access and storage which causes some flexing. If lightweight ceiling joists are used for an amateur loft conversion as a playroom or similar then the amount of flexing in the ceilings may be considerable and cracking to joint lines extensive.

A skim finish is not really sufficient for a plasterboard ceiling, and if crack lines are to be minimised then lining with a stout lining paper such as Anaglypta or woodchip is recommended before painting.

All internal surface coatings which are applied in successive layers, such as textured coatings over existing plaster, are only as durable as the material beneath and the correctness of the application. A temptation arises to use surface coatings as a means of covering soft and loose plaster with the result that the surfaces eventually laminate or shell away from the bases. During the course of any survey it is recommended that wall and ceiling plaster be checked for loose and hollow areas and particular attention be given to areas recently covered over with new coatings or decorations in circumstances where soft or loose plaster beneath may be suspected.

Condensation

Problems of decorative finish to internal surfaces will be exacerbated by the presence of excessive condensation. Condensation is a major source of complaint in post-War buildings, especially buildings with a poor level of thermal insulation to external walls and poor overall design. Condensation problems at their worst can make a building uninhabitable due to the mould growth which forms on wall and ceiling surfaces and inside cold cupboards. This problem was virtually unknown in pre-War construction and has resulted from a combination of factors including changes in living patterns and construction methods.

Causes

In order to achieve a satisfactory surface finish, the effects of dampness and condensation must be dealt with at source. It is not sufficient to cover over problem areas which will subsequently reappear. Dampness to wall and ceiling surfaces may arise either from one or any combination of five possible causes, which are:

(1) Direct weather penetration from the outside;
(2) Leaking rainwater equipment and overflows causing a point of penetration;
(3) Plumbing leaks within walls, behind ceilings or in adjoining areas;
(4) Rising dampness originating in the site;
(5) Condensation.

It is necessary to distinguish dampness created by (1) to (4) and involving specific remedies from condensation problems which are more complex and difficult to eradicate. Condensation occurs when damp air comes into contact with cold surfaces. Put at its simplest, the remedy lies in raising the temperatures of the wall and ceiling surfaces to above dew point. Alternatively, the amount of water vapour in the air may be reduced. Steps could be taken both to raise the surface temperatures and also reduce the amount of water vapour in the air. By such means condensation would be eliminated.

Remedies

Before dealing with the means by which surface temperatures can be raised it is necessary to discuss the more obvious causes of high water vapour content in the air. Certain simple precautions may be taken by the occupier of a building to reduce condensation levels by adjustments to living habits. Paraffin or flueless gas heaters discharging the waste products of combustion directly into the air within a building are a major source of condensation, and a change in the method of heating could be all that is required to eliminate the problem. Since the burning of hydrocarbons in oxygen produces large quantities of water vapour, paraffin heaters and flueless gas heaters of various types are really unsuitable in many situations and decorations are often affected, even on well insulated and warm wall and ceiling surfaces. Frequently windows will be found to be running with condensation where a paraffin heater or flueless gas heater has been in use for some time. One litre of paraffin produces about one litre of water in combustion. Experience tends to indicate that this type of heating is used intermittently, often in older buildings and frequently by occupiers in lower income groups for economy. Intermittent heating is itself a cause of problems since the air within the structure is warmed but the structure itself remains cold, an ideal condensation-forming situation.

Kitchens and bathrooms are obvious sources of water vapour, and when these rooms are in use their doors should be closed and good ventilation provided. The amount of steam in a bathroom can be reduced by the simple precaution of putting a little cold water in the bath first before running the hot water tap rather than *vice versa*. Appliances such as tumble dryers should always be provided with ducting to the outside air and damp washing should not be dried indoors in front of fires or over radiators. By means of small changes in heating methods and living habits, the amount of water vapour generated within a building may be substantially reduced and in a survey report a client could be advised accordingly.

Having dealt with the question of high water content in the air and the means by which it can be reduced, a condensation problem may still remain and this invariably arises from the bad design of the building, a matter about which a client will need to be informed. Poor design can arise from the use of unsuitable materials, from an illogical accommodation layout or from lack of attention to thermal insulation and ventilation.

The thermal insulation standard to apply as a suitable measure giving the performance of walls and ceilings should be that prescribed by current Building Regulations, and if that standard in a particular building falls short then it is recommended that this be explained to the client. Thermal insulation standards have risen appreciably in recent years so that many older buildings will fall well short of modern requirements. However, it should not be assumed that this fact will be accepted unquestioningly by clients. A client moving from a modern house with cavity walls and an insulation block inner skin, perhaps with a cavity filling and 100 mm of fibreglass quilt over first-floor ceilings, will need to be warned about the much lower standards which may apply in a house built inter-War with solid 225-mm brick walls and negligible ceiling insulation. In particular, the walls in bathrooms and kitchens in such a house may be prone to surface condensation, especially if the client keeps all the windows shut, takes frequent hot baths and showers and uses mobile gas heaters for heating. Such a client will not take kindly to the resulting peeling wallpaper and mould growth in the back of fitted wardrobes fixed to the external walls.

A note should be taken of the accommodation layout. For maximum efficiency, the greatest volume is enclosed by the least area of outside wall in a two-storey building in the shape of a cube (discounting curved walls). It follows that heat losses in single-storey buildings are higher and heat losses are also higher from projecting bays and wings. If a kitchen or bathroom forms a projecting wing to a building, perhaps on one floor only, then it may have three external walls plus a roof and a consequentially high level of heat loss. It will often be found that large old houses converted into flats will have small wings added to provide bathrooms or kitchens, or these may have been converted out of existing sculleries and outhouses with the original solid walls retained. In such circumstances these rooms may be cold, damp and ideal condensation traps.

The level of insulation above ceilings which are under flat roofs or balconies can rarely be checked without special exposure. In general, the heat losses associated with flat roofs and balconies are higher than would be the case where conventional pitched roofs are used. The interior of pitched roofs can normally be inspected and additional insulation may be provided without too much difficulty. The provision of additional insulation inside a flat roof is more difficult, necessitating either the lifting of the top surfaces or the removal of sections of ceiling. As a result, a number of methods have been devised for improving the thermal insulation performance of flat roofs, by the provision of either additional material above the decking or false ceilings beneath it, these then incorporating insulation.

In considering the design and thermal efficiency of a structure, small projections and areas of heat loss should not be overlooked. For example, small roofs to bays in older buildings often have no insulation at all. The floors of rooms located over integral garages, verandas and open porches should generally be separately insulated to reduce heat losses through those floors into the areas beneath. This is frequently overlooked.

Condensation is the likely cause if high moisture readings are obtained in wall plaster located on a cold external wall and readings are higher in the main and upper areas of the wall but lower at skirting level. Mould growth on wallpaper and within cupboards at high level is typical of condensation damage. Dampness from penetration is less general and related to specific exposed walls or points of penetration. Dampness from rising moisture will be concentrated at skirting level and will reduce in severity with wall height, rarely extending to higher levels above, say 1 m from damp-proof course level.

When mould growth on wall surfaces and inside cupboards is found this can be treated using fungicidal paints or fungicidal wallpaper adhesives. The use of fungicides to limit the development of moulds must, however, be regarded as a palliative rather than a cure and to be unsatisfactory in the long run. It is clearly preferable that the basic underlying causes of mould growths be dealt with by seeking out the sources of excessive water vapour or defective design leading to cold internal surfaces.

One useful means of raising surface insulation standards is to line the internal face of exterior walls with 3 mm polystyrene under wallpaper. This is particularly suitable in older residential property with solid walls and will help to compensate for the lack of a cavity. The disadvantage with this type of surface application is that the wallpaper and lining is prone to damage from denting. The overall thermal efficiency of the exterior walls is, however, much improved and condensation problems on walls are generally reduced or eliminated. When re-papering becomes necessary, the insulating lining usually has to be renewed at the same time.

Condensation on windows can be reduced appreciably by double glazing. Metal-framed windows are more prone to condensation difficulties than timber-framed, particularly the older type of single-glazed galvanised-steel casements much used in pre-War and nineteen-fifties construction. Often the condensation from windows will cause staining to the reveals inside and this may be incorrectly diagnosed as weather penetration around the frame. The polystyrene lining already mentioned is effective in preventing such staining to window reveals but again, it must be regarded as an expedient rather than as a permanent cure.

Tiled surfaces

Tiled surfaces are usually found in bathrooms, shower rooms, cloakrooms and kitchens. Good quality wall tiling should be plumb, level with all joints evenly proportioned and neatly grouted. It is a good idea carefully to examine wall tiling not only to inspect the standard of workmanship but also because the pattern of the tiles at wall junctions will give an indication of the evenness or otherwise of the wall surfaces and the extent to which walls may be out of plumb. Wall tiling, being inflexible, will show clearly the extent of any structural movement as accurately as any glass tell-tale so that gaps or cracks which have appeared in the tiling can indicate the extent of past structural deformation. Similarly, tile sills, internally or externally, may show signs of past separation at window openings.

The tiles should be tapped lightly all over to detect loose or hollow areas. The modern practice of tiling using spot adhesive is not always as durable as the traditional tiling applied to rendering encountered in older buildings. Older tiling will often be found to be very firmly fixed indeed and removal of tiles presents a problem. During renovation work it is often easier to re-tile over the original tiles rather than persevere with attempts to hack them off, and tiling over tiles can be satisfactory provided that the older surface glazing is cleaned and abraded first to provide an adequate key.

Tiling of various types is commonly used on floors. Over timber floors, the tiling should be capable of accepting flexing inherent in timber construction, while brittle tiles such as ceramic types, which require a firm solid base, are unsuitable for suspended timber floors. Vinyl asbestos tiles or vinyl sheet flooring can be laid over timber floors provided that an intervening layer of hardboard or plywood is provided and the degree of movement in the floor construction is minimal.

Exterior finishes

Exterior surfaces are often rendered and finished in pebbledash, spardash, smooth stucco, roughcast or stippled surfaces of various types. Such surface treatments must be suitable for the conditions of exposure in order to resist frost damage. They must also be designed to accommodate any movement in the construction beneath and to allow for the evaporation of trapped vapour.

In all cases the life expectancy of a rendered surface will only be as good as the adhesion to the surface beneath. Frequently, renderings are applied to walls which are in poor condition as a means of disguising the defects beneath and the key will be inadequate. In due course, the rendering will shell away from the base and the first signs of this, to be detectable by the surveyor, will be surface cracking and raised areas which are hollow underneath.

Various types of surface coatings are produced for painting roughcast and stucco renderings. Sprayed surface coatings are heavily promoted nowadays with various claims about the life of the treatment. As with the renderings themselves, the surface coatings will only be as durable as the base to which they are applied and the key is vital. Sprayed coatings are often used as a means of binding together loose or soft rendered surfaces and will not prove durable in such situations. Sprayed coatings are also employed as a means of limiting damp penetration through solid walls by providing an impervious coating and can actually be counter-productive in such a situation as, by preventing the escape of trapped moisture in the wall, they may increase a tendency for moisture evaporation from the internal surfaces where the barrier to penetration is incomplete or there is rising damp in the wall.

Some housing is given a new external facing of artificial stone on existing wall surfaces to alter the appearance of the building and (according to the manufacturers) improve weather-resistance and insulation. If such stone re-facing is encountered, certain basic questions need to be considered regarding the strength of adhesion to the base, the suitability of the base and any need to accommodate movement. Generally well-keyed brickwork or stone work in sound condition will provide a suitable base for re-facing, especially if shot-fired ties are used at suitable intervals. It is questionable, however, whether or not the re-facing of brickwork or stone in such conditions is worthwhile, discounting any aesthetic considerations. If brickwork or stonework is in poor condition and if re-facing is used for this reason, then this must call into question the long-term degree of adhesion between the new face and the base.

The application of traditional wall renderings requires degrees of skill, especially in the case of stucco which needs a careful selection of mix so as to be neither too lean nor too rich in cement. In such work, careful application is needed with a minimum of working of the surface. If stucco is overworked, the resultant surface will contain more finer particles than the backing and slight differences in the rates of expansion and contraction between surface and base will result in the familiar patterns of hairline cracking so often encountered in such a finish which has not been decorated recently.

Rendered exterior walls which are subject to constant water saturation from driving rain, leaking gutters, overflows and other causes, will be at risk from frost damage. South-west walls usually face the prevailing weather and will be most at risk. In parts of Wales, Scotland and the West of England where the degree of exposure is higher, it may be that all walls will suffer saturation from driving rain at some time. The crystallising into ice of the water between the rendering and the base will cause that rendering to fall away, sometimes large areas at a time.

Asbestos

In connection with finishes and surfaces in buildings special mention should be made of asbestos. Health risks associated with asbestos fibres are now well understood and during the course of any inspection surveyors should be careful to check whether asbestos has been used for cladding, pipe-lagging or other purposes; if necessary, and especially if the presence of blue asbestos is suspected, clients should be advised to obtain independent advice and analysis of any asbestos or asbestos-based materials.

Generally asbestos as used with cement to form roofing sheets, water tanks and the like is stable and unlikely to release fibres into the atmosphere in dangerous quantities unless drilled or broken; such stable materials can however become more friable with age so that any asbestos-cement found to be in a friable condition could be hazardous.

Blue asbestos is highly dangerous and should not be disturbed if found. It should be removed by specialist contractors using approved handling methods to avoid any release of fibres into the air. The material is unlikely to be found in private residences but will be encountered in some post-war and late pre-war industrial and institutional buildings, generally in the form of pipe and boiler lagging.

Chapter 9

Services

There are a number of things we are not happy about. For one thing the lavatories are flushing hot water.
New house purchaser

In all types of property, services will be a matter of major interest to any prospective purchaser. He will wish to know whether or not the services are functioning satisfactorily and if they are likely, in the future, to give trouble needing expensive repair. Some clients will be remarkably knowledgeable about services and others will be entirely ignorant and require considerable guidance, especially regarding the effectiveness of heating systems.

During the course of the survey, the inspection should include an examination of services at all points where they are visible. Great care should be exercised not to assume that services which appear modern have been renewed in their entirety. Amateur workmanship is rife in the areas of electric wiring, plumbing and heating; nothing should be taken on trust.

Electricity

The Institute of Electrical Engineers (IEE) Wiring Regulations recommend that all installations be periodically inspected and tested. For domestic residential buildings the recommended frequency is every ten years. Places of work are covered by the Electricity at Work Regulations 1989 which require periodic inspection and testing by law without stipulating precise frequencies. The signatories to electrical certificates should normally be one of the following:

(1) Chartered Electrical Engineer
(2) Contractor Member of the National Inspection Council for Electrical Installation Contractors (NICEIC) or the Electrical Contractors Association.

A surveyor undertaking a Building Survey will not be qualified to undertake any testing unless he, or she, is coincidentally also a member of one of these bodies. If the electrical installation is obviously old and the wiring clearly in need of renewal no electrical test is necessary since a surveyor can advise the client to budget for complete rewiring and obtain a quotation before buying.

In all other cases a test and electrical certificate would be advisable whenever a property changes hands and clients should be advised to have any work done which is necessary to comply with current IEE wiring regulations, the latest of which was the sixteenth edition published in 1991 and which in October 1992 became BS 7671:1992.

Modern PVC sheathed cable, used generally since about 1958, has proved durable and so far I have not encountered failures in PVC sheathed systems except where grossly overloaded. Tough rubber, vulcanised India-rubber (VIR), and lead-sheathed cable, used prior to about 1958, must all now be regarded as due for renewal. Rubber-sheathed cable has a safe life-span of between 20 and 25 years, tending to become brittle after this time, especially on power circuits used regularly for high-wattage appliances, and is liable to attack by gnawing mice in the case of older properties. Lead-sheathed cable can be mistaken for grey PVC cable, especially in, say, a dark roof void, but if a scrape with an insulated screwdriver leaves a bright shining line this will confirm the presence of lead sheathing.

The surveyor should be able to discover and comment upon certain obvious defects in electrical systems, and the following points should be checked:

1. At the service head, the overall condition of the switchgear and cable; generally a modern consumer unit with all circuits neatly labelled is to be hoped for, but invariably in older buildings a multiplicity of iron-clad or wooden boxes will be found and replacement of these by a modern consumer unit can be recommended. If rubber- or lead-sheathed cables are found, an assumption may be made that re-wiring will be required and the client should be advised to budget for this pending receipt of an electrician's report. It will be appreciated that ring mains used since the Second World War may themselves need re-wiring if rubber cable were used (mostly this applies to installations prior to 1958). Clients often believe that a ring main installation indicates that the system is modern and will therefore not need attention, but this is not always so.
2. The visible power sockets, light switches, pendants and fixed electrical appliances should be examined. For safety reasons cord switches are required in bathrooms and shower rooms where water is present and within certain distances of the switch. Immersion heaters should generally be on a separate 15-amp circuit with double-pole switch and have an indicator light. Such heaters on plugs and sockets, or spurred off a ring main, may not be satisfactory. A check should be made for frayed cables to immersion heaters and other appliances. Nowadays lighting pendants are earthed but this has not always been so. If the pendants are not earthed this may not, of itself, justify action but a client's attention should be drawn to the point. Unswitched power sockets are permitted by regulations but switched sockets are preferable. If unswitched sockets are found, clients may be told that these could be replaced with double or single switched sockets if required. Fixed electric wall heaters should be on 15-amp circuits with fused isolating switches and in bathrooms or shower rooms should be cord-operated and mounted high enough that they cannot be touched accidentally.
3. The visible cable on walls and in voids should be examined. Cable clipped to walls below a height of 1.5 m should be in conduit so as to be protected from damage. Unprotected surface-clipped cable run along skirtings is not satisfactory. In roof-spaces the lighting cables should be taken to junction boxes, and although neat clipping is not essential it is desirable, particularly if the area is likely to be used for storage or occasional access. Wiring joined with insulation tape is unacceptable.

If the system is modern but obvious breaches of the Institution of Electrical Engineers' Regulations are found, or the installation has not been made in accordance with good practice, then the surveyor should be on his guard over the possibility of sub-standard or amateur work and advise a test before the system is used.

If the client wishes to cook with electricity, a separate circuit with 30-amp, 45-amp or 60-amp fuse should be recommended with a suitable grade of cable, depending upon the nature of the appliance to be used. A combined cooker panel and power socket is normally provided on the kitchen wall. It is essential that cookers are not run off ring mains, since this will result in overloading.

Any external circuits have to meet with special requirements and be suitably encased in watertight conduits, whether below ground or above. Exterior wiring is often amateur and, if so, may be dangerous. A particular point should be made of having any such wiring checked for safety.

Gas

A surveyor should recommend that the gas service, together with any fixed gas appliances included in the sale, be inspected and tested for safety as a normal routine precaution when the property changes hands and before any gas is used. The inspection and testing must be undertaken by a CORGI registered contractor. Having given this advice the surveyor may also make any other comments which result from his own inspection.

The writer makes a point of examining the gas service head and meter, and if any gas smell is detected (a not uncommon occurrence) then this should be brought to the client's attention. If the building is old there may be a considerable amount of gas pipework running under the floorboards. It is not unusual to find original pipework for gas fires and gas lighting still in place and although this may be capped off where appliances were removed, it may still be connected to the service and at mains gas pressures. Older installations were designed for town gas which was not only supplied at lower pressures than modern natural gas but also contained more moisture, and this assisted in maintaining seals on threaded pipe joints. The dryer natural gas at higher pressures will often find weaknesses in such old pipework and leaks often result.

Occasionally, gas pipework will be found to be heavily rusted and for this reason it may be on the point of failing. Underground gas service pipework often leaks and requires renewal if old, especially if the ground in which it is laid is subject to vehicular traffic. The surveyor may mention this possibility and point out that any such underground pipework will not have been inspected as part of his survey.

A gas leak under the front garden of a house usually seems to occur just after some expensive re-paving and re-turfing of the front garden has been carried out – with adverse effects on the client's blood pressure!

Ventilation to conventional flued gas appliances is a matter of the utmost importance and should always be checked. Generally, for safe running a conventional flue boiler or warm air heat exchanger unit must have ventilation of the equivalent of a 225 mm × 150 mm airbrick to the outside air at all times. The increasing tendency towards improved insulation and draught-proofing of houses means that the structure becomes virtually airtight and if no ventilation for a conventional flue is provided, this could result in the suffocation of the occupants by removal of oxygen from the atmosphere.

A careful check of ventilation provision must be made and a warning given to the client not to alter or block ventilation in any way. Balanced flue appliances do not take oxygen from the room in which they are located and no requirement for special ventilation provision arises in this case.

Boiler location and other matters are governed by the Gas Safety (Installations and Use) Regulations 1984 which prohibit, amongst other things, the location of any flue under or adjoining any opening window. Examples of unsuitable flue locations will often be found during survey and should be mentioned to the client with suitable advice.

Certain small flueless gas room heaters have been available in the past and are still sold. This type of heater, like paraffin heaters, generates considerable vapour during combustion, and good ventilation for their safe use is essential. With this type of heating deterioration of decorations invariably results, even where there is good ventilation. If ventilation is poor and the fabric of the building cold, serious condensation problems occur leading to mould growth on walls and in cupboards.

Water

The plumbing for the hot and cold water supply must be examined where this is accessible. The location of both the company stopcock and the internal stopcock should be confirmed and if there is no sign of these the client must be advised to have them traced and checked.

Lead plumbing is still often found in older buildings and even if internal plumbing is modern, an original lead incoming and rising main supply pipe may have been retained. For health reasons lead pipes are no longer recommended. In the past it was believed that the lead could not enter the water in the system due to the protection inside the pipes provided by the coating which forms when the surface of the lead oxidises. This is no longer considered sufficient protection since the health risk which arises does so from a total long-term build-up of lead in the body from a variety of sources of which lead plumbing may be only one. Accordingly, while it would be foolish for the surveyor to be alarmist on this topic, if lead pipes are found or any lead-lined tanks, the appropriate advice must be to recommend their removal and the substitution of polythene, copper or similar modern materials.

During the course of the survey, the accessible plumbing and all sanitary ware, taps, cisterns and waste pipes are visually inspected and tested by normal operation. The surveyor should note any weeping or corroded joints, dripping taps needing new washers and noisy plumbing due to water-hammer or other causes as well as running overflows and the consequent possibility, if header-tank ball taps are found to operate correctly, of a leaking coil in an indirect cylinder.

Occasionally in an older building, the plumbing may have been turned off at the stopcock and the overall age and condition of the system may be such that re-charging the system with water would be ill-advised. In such circumstances it would be better not to charge up the system for tests but to advise the client generally that having regard to the obvious defects in the system, it should not be used until inspected by a plumber or perhaps renewed. Old stopcocks may be difficult to turn and can even seize or break off after they have been opened.

Considerable damage could be caused then to a property if defective plumbing is charged up in circumstances where it may be difficult to turn off and drain it down again quickly.

It is a good idea to check the fixings to wash basins as these are often found to be loose. The caulking to bath edges is important and it is difficult to maintain a good seal with acrylic baths due to their flexibility. Caulking around showers and baths with shower attachments is essential and a watch should be kept for any points where a regular spillage of shower water may have taken place with possible adverse effects on timbers or ceilings beneath. Water-closet waste-pipe and flush-pipe seals are often found to be weeping at the back of the pan and any long-standing leak in this area may have resulted in deterioration to the floor or ceiling beneath. This often arises if the pan is not mounted exactly square to the cistern, or if modern press-fit joints have been wrongly used, or if the pan is mounted on an insufficiently rigid timber joist and boarded floor.

Certain combinations of metals in plumbing systems can give rise to galvanic corrosion. Galvanised-steel cisterns and hot tanks, for example, will rust through rapidly in the presence of a lead and copper piping. A sacrificial anode consisting of an aluminium block earthed lead to the steel tank may extend its life although all galvanised-steel cisterns and hot tanks will rust through in time, whether protected or not, and replacement with a modern plastic or glass-fibre type is usually necessary once rusting has set in.

Central heating

Fig. 9.1 illustrates diagrammatically the layout of a typical water- and central-heating system likely to be installed in a small residential or commercial property dating from the nineteen-sixties or nineteen-seventies. Such a system is simple with only a single pump on the heating circulation and very little to go wrong. Provided that the system is correctly installed and the radiators and pipework are protected from chemical corrosion by an inhibitor, such a system should prove reliable and serviceable for 20 to 30 years before major renewals or repairs are needed. Occasional servicing and attention to the boiler and pump may be expected and a minimum of annual maintenance is recommended before each heating season.

Certain elementary principles apply in such a system. Overflows should be taken separately to the eaves and the header overflow should not be run into the main cistern. All expansion pipes should be of greater diameter than corresponding feed pipes and an overflow pipe diameter should be greater than that of the corresponding supply pipe. Cylinders must be indirect otherwise a mixing of circulation water and hot service water will occur. It is essential that this should not occur as the addition of fresh water into the circulation will add oxygen and cause rapid corrosion to steel radiators and fittings. This may happen where the coil in an indirect cylinder has a split or pinhole and can be confirmed, as previously explained, by a continuous running of the system's overflow.

Occasionally it is puzzling to find an indirect system which does not have a separate header tank. This invariably will be a system with a self-priming cylinder, commonly used in the nineteen-fifties and nineteen-sixties, and can be confirmed if the cylinder has a 'Primatic' label. Self-priming cylinders employ a one-way valve to permit the addition of water from hot into indirect service. In self-priming systems,

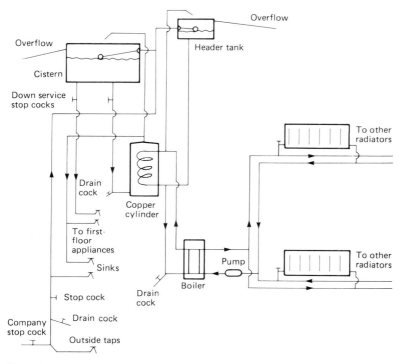

Figure 9.1 Diagrammatic layout of a simple domestic water and heating system using copper piping in 15, 22 and 28-mm diameters. Indirect gravity primary circulation is provided to the cylinder. A pumped heating system is provided in small-bore, two-pipe layout to pressed-steel panel radiators

defects can easily arise involving a mixing of the water and the lack of an expansion header tank means that inhibitors and corrosion proofing chemicals cannot be added to the system. A separate header tank and indirect cylinder are much to be preferred as seals have a habit of breaking down which can introduce rusty radiator water into the hot water service.

Some central heating systems are found to be laid out as a one-pipe rather than a two-pipe layout. In one-pipe systems, all circulating water enters and leaves every radiator via a single pipe, usually of wider diameter than normal small-bore or micro-bore. One-pipe systems can be gravity fed or pumped. In either case they suffer the disadvantage that the circulating hot water cools progressively as it runs through the radiators so that the radiators nearer the end of the circulation in a pipe run are colder than the earlier ones. Traditionally the progressively cooler running can be compensated by employing correspondingly larger radiators towards the end of long pipe runs but this is of limited use when radiators are turned off. One-pipe systems are thus generally difficult to balance.

Gravity heating circulation uses wide-bore pipework of either copper or iron and is used in many older buildings including houses, blocks of flats and commercial structures of one kind or another. Gravity circulation is more sluggish than pumped circulation and older gravity systems inevitably contain a considerable sludge build-up with furring in pipework which further restricts the flow. A specialist heating engineer's report on older systems is inevitably necessary in order to confirm whether or not they are still functioning with reasonable efficiency.

All tanks and plumbing in lofts, under floors and at other points where frost damage is possible should be well lagged and tanks should always be covered. Various unsavoury items may be found in uncovered tanks including dead birds and, apart from the health hazard posed, there is a real danger of blockages occurring in the pipes and overflow outlet.

System control

In current domestic central-heating systems a system of control and layout may be adopted which is more sophisticated than the simple system illustrated in *Fig. 9.1.*

A fully-pumped system could be employed using either two pumps, one for heating and one for hot water, or a single pump and two motorised valves, or a single pump and motorised valve giving heating or hot-water priority. The advantage of fully-pumped systems is that the programmes for hot water and central heating may be quite separate, giving greater flexibility of control, and cylinder and radiator temperatures may be separately set using a strap-on thermostat to the hot water cylinder.

A secondary advantage of the fully-pumped system is that the boiler does not need to lie close to and below the level of the cylinder as is necessary for gravity circulation. The cylinder warm-up is also more rapid than with a gravity feed.

In addition to the use of a normal wall thermostat for setting room temperatures, systems of zone thermostats can be employed sited either floor-by-floor in a building or as thermostats controlling individual radiator valves. Generally speaking, the more sophisticated the system becomes, the more there is to go wrong and a point is reached beyond which the advantages of sophistication are outweighed by the disadvantages of complexity, including the cost and inconvenience of additional maintenance and repair.

Small-bore systems have, in some cases, given way to micro-bore, especially where central heating is installed in an older building previously not so heated. Micro-bore (or mini-bore) is easier to install since pipework is flexible and more easily concealed. Most micro-bore systems will incorporate some small-bore pipework around the boiler with the main pipe runs in micro-bore taken from central manifolds and either run under the floors or boxed in. Such systems require careful design to ensure that the small pipework is adequate for the velocity of circulation required to achieve the desired room temperatures.

Heat emissions from a boiler casing are equal to those from a small radiator so that if a boiler is located in a small room, say a kitchen or utility room, the benefit of heat emissions may be valuable in winter-time, making the overall performance more efficient. Often boilers are found to be located in exterior boiler houses, in which case the benefits of heat emissions from the boiler casing are lost on the outside air. In domestic installations the use of external boiler housing appears to be a practice dating from a time when fuel costs were low before energy costs began their present steep rise and the overall efficiency of the system and prevention of heat loss were not regarded as being so important.

The controls for most systems incorporate timed switching. A programmer is preferred with a selection of timed programmes so that heating systems may be left to run automatically as required. In a residential property, a fully-pumped small-bore system with two motorised valves is ideal with a strap-on thermostat regulating cylinder temperature and a wall thermostat in the main living room regulating radiator temperature. Such a system may be either gas- or oil-fired. Under regulations which came into force in July 1987 continental-style unvented heated systems are now permitted in the UK. These systems have a pressurised boiler and there are no storage tanks.

In addition to conventional water-filled central-heating systems a wide variety of other methods of heating may be encountered which could form part of the structure under review, or alternatively are chattels, including various types of free-standing gas and electric fires, electric storage heaters, mobile gas and paraffin heaters and solid fuel appliances including wood-burning stoves. As such items are chattels included with the property it is recommended that clients be advised to satisfy themselves on the condition of appliances and their performance, including their running costs. Heating requirements are generally somewhat personal and a means of heating which will be satisfactory for one client may be unacceptable to another.

Flues

In the case of heating systems which use flues it is recommended that clients should be advised that flues must be swept and checked as a normal routine when a building changes hands. A build-up of soot in a flue can cause a chimney fire and this can be a hazard in an old building where a flue lining may be poor or non-existent. A risk can arise of timbers having been built into chimney walls, especially at roof level, this giving a combustible access for a chimney fire to spread into a roof space. Many cases have occurred in old buildings of beams or purlins having actually been built into a flue and smouldering for perhaps months before bursting into flame with disastrous results.

A cautionary note should be sounded over flues from wood-burning stoves which are often installed in existing fireplaces. Manufacturers of these appliances will often recommend that the dampers should be opened every 24 hours and used at maximum heat to burn off the tar deposits which accumulate in the flue. The temperature of the flue linings can reach 850°C, and if tar ignites it could reach 1500°C, and serious fires have occurred as a result. Moreover, chemicals are sold to the owners of wood-burning stoves to assist in dissipating flue deposits. Chemicals using phosphorous cause intense heat and those including chlorine can result in chloride attack to linings, especially old parging and brickwork.

If roof and floor timbers are in contact with the walls of flues or even built into these walls (a not uncommon fault in older buildings), they are likely to be subject to intense heat to the point of combustion by this practice of flue tar burning. Only flue linings of a very high standard will resist such aggressive conditions and when a wood-burning stove is encountered, it is suggested that the suitability or otherwise of the flue arrangements should be reviewed critically.

Drainage

Stormwater

Stormwater drains generally consist of gutters and rainwater pipes which discharge into either open or sealed gullies and with surfacewater gullies provided to paved areas. The stormwater may run to soakaways or to a separate stormwater sewer or into a combined foulwater and stormwater sewer. It is recommended that the surveyor should familiarise himself with the accepted practices in the locality concerned as local water authority regulations may provide, for example, that stormwater should not be run into the foul sewer, especially in inland locations where large quantities of stormwater would overload the foul system. Soakaways are suitable in porous, well-drained soils but unsuitable for heavy clay or non-porous soils which drain reluctantly, unless there is no alternative.

Frequently the surveyor will find the storm drainage system to be a mystery. There will often be no manholes nor any indication of the drain runs. Soakaways tend to block eventually and the pipework becomes choked, especially if leaves fall in the gutters and no cages are provided over rainwater pipes. Soakaway systems often need to be relaid after a period of time as they cannot easily be rodded through.

The importance of avoiding point discharges of stormwater into the soil around foundations has already been stressed in Chapter 3. In many cases the surveyor should consider whether or not the storm system ought to be exposed and checked. Perhaps the best guide to this is to suggest that a client checks the system during a period of really heavy rain to see if the storm gullies allow a free flow. If they are blocked it is likely that stormwater is draining into the soil around the building from overflowing collars and gullies, and this is undesirable. A good modern practice is to lay storm drains watertight for a distance of at least 3 m from the main walls and thereafter in land drains to a soakaway if required. Some intermediate rodding access is recommended in the form of small rodding chambers over any bends. If the storm drains run into the foulwater sewer or to a separate storm sewer, a watertight system of a standard similar to the fourwater system is required with suitable rodding access.

Foulwater

Foulwater drains should always be provided with adequate means of access for rodding and attending to blockages, and provision for rodding at each change in direction and each branch drain connection is necessary. If this is lacking, it is suggested that this be mentioned in the report together with suitable advice. Lack of good rodding access on a branch drain from a gully may not be critical but lack of access to a branch drain from a WC or soilpipe serving a WC is unacceptable if a good standard of design is to be maintained.

Foulwater drains will normally take the form of either a public sewer or a private sewer or a private drain. These distinctions are important since they affect the client's potential liability for drain repairs both within and outside his own curtilage.

Sewers

A public sewer is maintainable by the local water authority or its agents and will normally be laid under a public highway or grass verge but can also run under private land. Generally, defects in a public sewer will not be the liability of the adjoining property owners connected to it once it has been adopted and it is not normal practice during the course of a building survey to attempt any examination or testing of public sewers serving the property under review. In some localities the public sewers, both for stormwater and foulwater, may become overloaded from time to time, and if the surveyor is aware of any such problem it would, of course, be open to him to discuss it in his report. The ability of sewers to cope with storm- and foulwater discharges may be relevant when changes of use or more intensive use are proposed. Changes of use normally require local authority permission and it is possible for consent to be refused if the existing public sewers are inadequate. Such cases are, if anything, on the increase, particularly in urban areas.

A private sewer is a shared drainage system serving a number of properties. The Public Health Acts provide that such drains are the joint responsibility of all property owners connected. In order to be able to confirm the extent of a client's potential liability for a private sewer, it would be necessary to inspect and possibly test the whole system. Clearly this is often impracticable; frequently 20 or more houses will be connected to a private sewer and a surveyor could find himself testing half the drains in a street in order to cover that potential liability. Where the subject property is connected to a private sewer, it is suggested that the significance of this be explained to the client and that he be given the opportunity to decide how far he wishes to pursue the matter.

In addition to the Public Health Acts which impose a liability for the maintenance of private sewers which can be enforced by the local authority, the question of responsibility for shared drains may also be covered in title covenants for freehold property, or lease covenants for leasehold property, or their Scottish equivalents. If the title or lease specifically identifies drains for which a client may be liable, these may require special attention from the surveyor, possibly needing access to other parts of a building or adjoining properties.

It is rare for a local authority to serve notices under the Public Health Acts requiring repairs to a private sewer but this situation does occasionally occur. It is clearly a matter of concern to a client buying a property to find that he may have to pay a contribution for some drain repair, perhaps some distance from his own property and about which he may be unaware until he is served with a formal notice. It is debatable whether or not all property owners connected to a private sewer must contribute or only those 'upstream' from the defect, as local authority practice in serving notices under the Public Health Acts appears to vary.

A private drain is a system which is exclusive to a particular property and runs through its curtilage and into a public sewer. Up to the point of connection to the public sewer the private drain is entirely the responsibility of the property owner (or lessee) concerned. Testing procedures for private drains are clearly much simpler than for private sewers.

Testing and maintenance

Two main forms of drains test are normally used by a surveyor, these being water-pressure tests (particularly suitable for traditional pipework between open manholes) and smoke tests (particularly suitable for above-ground soil stack systems). Drain testing procedures and the various types of drainage systems are well documented in the building text books and are not discussed in detail in this volume. However, certain comments on the interpretation of test results would seem worth stating.

The first point to bear in mind is that the standard of test result required for new drainwork will rarely be achieved in existing drains. Old drains may well be watertight in normal use when running perhaps one-third full and, as such, may be perfectly serviceable. A water-pressure test to any height will impose abnormal pressures on the pipework and manhole benching and will often indicate leakages. Such leakages may not of themselves justify the relaying or pressure-grouting of the pipework, and the degree of leakage, the invert depths and the general condition of the system should be taken into consideration when deciding whether or not major repairs are justified.

It is an undoubted fact that a majority of pre-War drainage systems will leak under test if the head of water used is high enough, but many will flow freely and be otherwise quite serviceable under normal conditions.

It is perhaps worth considering that a client might expect from his drainage system:

(1) That the drains will flow freely in normal use without blockage.
(2) That in the event of blockage occurring it can readily be cleared.
(3) That the system be airtight at ground level and within the building so that no drain smells or health hazards arise.

Most common blockages are caused by stepped fractures in pipework, tree-root intrusion or rough workmanship to joints and manhole benching. These matters will often be detected on a visual examination within manholes and through pipework using a mirror. Shallow drains running under or close to trees should always be checked for tree-root intrusion, even if the test result is reasonable, since the fine root tails can enter the drains through narrow fissures and even effectively seal them off in some cases.

If the drain gradient is too shallow this may cause blockages and rough checks on the invert depths will generally indicate whether or not the gradient is too shallow. If a level is not available, a check can often be made on drains around a building by taking invert depths in relation to the main damp-proof course line of adjoining walls (assuming damp-proof courses are not hidden).

Manhole covers will, more often than not, need cleaning, greasing and re-setting to provide an airtight fit and cracked manhole covers should always be renewed to keep them airtight. Loose frames should be re-set in new cement haunching for the same reason.

In some districts fresh-air inlets in conjunction with interceptor traps were used on private drains in the past but may no longer be regarded as necessary where more than one vent pipe is provided. When this is so, we generally suggest that fresh-air inlets be sealed airtight in cases where other ventilation is to a suitable modern standard. The exception is where an interceptor trap is fitted in the last manhole before the sewer and a fresh-air inlet with mica flap valve ventilates the drain in conjunction with a single vent pipe at the head. Mica flap valves in a fresh-air inlet to keep out vermin are often found to be broken or missing.

Some modern uPVC underground drain systems, such as the Marley rodding point system, dispense with traditional manhole construction for most purposes and instead provide angled rodding eyes at ground level for each change of direction. Testing using drain plugs or inflatable bags is still possible but a visual examination is not possible through the pipework without recourse to video equipment and this is not usually suitable for the narrowest pipe diameters.

In rural areas, private drainage systems encountered vary from the modern, efficient and self-cleansing to the ancient and insanitary. The writers have encountered drains with no access whatsoever and no indication of where they terminate, also drains discharging directly into ditches and watercourses or into bore-holes. For example, in the Mendip Hills, house drainage systems used to discharge into pot-holes known locally as 'swalletts' which eventually directed the effluent into the cave system beneath. Clearly such primitive systems are often highly unsuitable and advice may be given that a modern system with ample access for cleansing be provided instead.

Cesspits

Cesspits are watertight sealed tanks which are periodically emptied by the local authority, their agents or specialist contractors and subject to emptying charges. If possible, the capacity of the tank and the amount currently charged for emptying should be checked. Clients may then consider if the tank capacity is adequate and the arrangements for emptying suitable. It is common practice for local authorities to insist on watertight cesspits and for these to be built in 225 mm brickwork, rendered both sides. The writer has heard of occasions when empty milk bottles would be built furtively into the brickwork and rendered over so that when the cesspit had passed final inspection, the ends of the bottles would be broken out thus permitting the tank to drain into the surrounding subsoil and reducing the frequency and cost of emptying. Rustic cunning of this type might be expected if the cesspit is required to be emptied only at infrequent intervals. The local authority or their agents will generally keep a record of emptying dates for all cesspits in their area and a telephone call can often elicit details of the number of emptyings over the last few years together with the amounts emptied.

It is essential that cesspits be located within range of the suction pipe from an emptying vehicle and that a parking area for the tanker is provided with suitable access from the highway. If there is any doubt about the correct location of the cesspit, this point should be checked with the local authority.

Septic tanks

Septic tank systems are well described and documented in the textbooks and I will omit any technical details of these. If a report is required on such a system, the following points should be borne in mind:

1. Septic tanks, like cesspits, also need emptying periodically. The frequency of emptying is, however, much less with a septic tank, perhaps annually or even every two years, depending very much on the design, capacity and the amount of solid material with which it has to cope and consequently, the amount of humus which accumulates in the bottom of the tank. Accordingly highway access for a tanker with a parking area within suction range is again necessary.
2. It will often be found that filters and soakaways become choked after a time and land drains sometimes need to be relaid if the system is old. Any evidence of a tendency for the system to block or flood may indicate choked soakaways. This depends to some extent on the length of the soakaway pipes, the lie of the land and the draining capacity of the soil. Gravel or chalk soils drain very well whereas heavy clays drain badly and are not suitable for taking soakaway systems unless considerably extended.
3. If there is any doubt about the design of the system or its suitability for the client's needs, a specialist drainage engineer's report could be suggested and the client given the opportunity to commission such a report before he buys the

property. Always lift the covers to the tank itself and look for the 'crust' on the surface which indicates that the septic tank is working correctly.

To sum up, private drainage systems are inevitably less satisfactory from a client's point of view than connection to a public sewer. If a public sewer is available and the cost of connection is not prohibitive, then the option of connection should be considered and, if necessary, recommended in the report.

Drains generally

The RICS Practice Note 'Structural Surveys of Residential Property' recommends that the surveyor undertaking a structural survey should open all accessible manhole covers, record the routes taken by the pipework below ground and observe the drain flow; also that the means of disposal of foul water and surface water be confirmed (whether by main sewer or otherwise) and that storm drains be tested with water to identify blocked gullies etc.

The Practice Note does not require that drains testing be undertaken as a standard or automatic procedure but that the surveyor should note features which might justify further investigation or tests. Such features might be evidence of past blockages, shallow drains running under paving which has cracked or subsided, drains running under trees, older traditional earthenware pipework running in shrinkable clays or drains with cracked channels and damaged benching within manholes. If it proves impossible to see between manholes using a mirror this may well indicate fractured or distorted pipework.

Pitch fibre pipes

Occasionally pitch fibre pipes may be encountered. Typically black in colour in contrast to the red earthenware or the orange or grey PVC normally seen, these were used for a short time in the post-war period during the 1960s but were found to suffer certain drawbacks. The pipes can deform under loading from above and were prone to damage from vermin since determined rats could gnaw through them.

If pitch fibre pipes are found these will generally need to be renewed. A pressure grouting repair is not possible and the drain has to be relaid. This is potentially expensive for clients and unless the nature of the pipework is confirmed on survey the problem may only become apparent when leaks or blockages occur and a cctv survey is undertaken.

Chapter 10

The report

The defendant failed to exercise the reasonable care and skill of the prudent surveyor which he had impliedly promised to exercise when he received instructions to survey the house. He committed a breach of contract with the plaintiff and provided him ultimately with a report which was so misleading as to be valueless.
Judge Kenneth Jones – Morgan v Perry (1973)

The surveyor's report should always be made in writing and be prepared with great care so that nothing intended to be included is omitted by mistake. It should be typed neatly on good quality paper and bound with a reasonably durable cover.

Paragraphs in the report should be numbered as well as the pages, or alternatively, sub-titles used so that in the event of discussion between the surveyor and his client either by telephone or otherwise, each party may readily refer the other to the section concerned.

The report should have regard not only to the particular circumstances and condition of the property under review but also the circumstances of the client. The preparation of a surveyor's report is a personal service between two parties and the service needs to be tailored to meet the needs of the consumer.

All surplus comment, padding and attempts at humour must be avoided. Few clients would feel happy about paying a fee for anything other than facts and information on the basis of which they will be taking an important personal and financial decision — possibly the biggest personal financial decision in their whole lives.

At the conclusion of the report it should be signed by the surveyor giving his or her name and qualifications.

Standard headings and clauses

In order to assist in drafting and in order to make the report intelligible to the reader, it should be divided by a series of main headings and sub-headings, the following list of headings and sub-headings being ideal for most situations although practice will naturally vary from one firm to another and in differing circumstances. From time to time a surveyor will be faced with the drafting of a report in circumstances where an entirely different layout from that normally used will be necessary, for example, in the structural report for a church quinquennial survey.

A suggested framework for a typical Building Survey report could be as follows:

1. Introduction

2. Conditions of engagement, limitations and exclusion clauses

3. Date of inspection and weather

4. Conclusion and summary

5. Legal matters

6. General description

7. Externally
   (1) Roofs
   (2) Chimneys, flashings and soakers
   (3) Parapets, parapet gutters and valley gutters
   (4) Gutters, downpipes and gullies
   (5) Main walls and foundations
   (6) Damp proof courses and sub-floor ventilation
   (7) Joinery and woodwork
   (8) Decorations and finishes

8. Internally
   (1) Roof voids
   (2) Ceilings, walls and partitions
   (3) Fireplaces, flues and chimney breasts
   (4) Floors
   (5) Dampness
   (6) Joinery and woodwork
   (7) Decorations
   (8) Cellars and vaults
   (9) Rot beetle and other timber defects
   (10) Thermal insulation

9. Services
   (1) Electrical
   (2) Gas
   (3) Cold water plumbing and sanitary fittings
   (4) Hot water and space heating
   (5) Drainage
   (6) Other

10. General
   (1) Garages and outbuildings
   (2) The site and boundaries
   (3) National and local legislation and regulations

11. Advice on value (if required)

At each stage the surveyor should make clear what has been inspected or tested and what has not been inspected and tested during the survey. No assumptions should be made regarding the condition of any hidden areas or untested services.

In inspecting a building for a potential purchaser in the first instance a surveyor should not feel obliged to expose and inspect any of the foundations, undertake trial bores or conduct other detailed investigations of the site. If site features encountered during the survey lead the surveyor to think that foundation exposure or site testing would be advisable then he or she should explain the reasons and thus give the client the opportunity to decide what to do.

Another example arises in the case of hidden floor construction. A suspended timber kitchen floor may be encountered having a covering of vinyl tiles on hardboard. A washing machine may have been leaking over this floor for some time and an inspection of the floor boards, joists and sleeper plates beneath could entail considerable disturbance and damage to the surface tiling. The surveyor is not expected to undertake such exposure during the initial survey but he or she should point out the circumstancs to the client. This then gives the client the opportunity to have the necessary exposure carried out, subject to a builder's attendance and the vendor's agreement.

The most effective way to assemble a report is to have a folder available containing all the standard report headings, sub-headings and clauses normally used by the surveyor's practice. This folder should contain a list of all the items which are to be covered in the report so that nothing may be overlooked. The information may also be stored on disk if a word processor is used.

In Chapter 18 a specimen report is included which could, if desired, be adapted and enlarged by the practitioner to form the basis of standard headings and clauses. Such clauses may also be fed into a word processor and used in this way to reduce repetitive typing work.

Standard clauses will not cover more than perhaps half the total report at the most, and whenever they are used care must be taken to ensure that they are fully appropriate in the circumstances. Unless appropriate to the situation in all respects they should be discarded and alternative wording used.

It is essential that standard headings and clauses be kept up-to-date and that it be continuously reviewed and re-written to keep abreast of current trends. In a partnership practice, periodic meetings of the partners are essential to discuss their own reports and possible improvement in their format so that standards are maintained.

Presentation of the report

The following, extracted from *The Chartered Surveyor* (April 1981), indicates one view taken by the Royal Institution of Chartered Surveyors' Professional Practice Committee:

"The Professional Practice Committee recently considered a structural survey report provided by a member and was much concerned by the careless way in which the report had been prepared with regard to spelling, grammar and technical understanding.

The Committee agreed that the report, as presented, prejudiced the professional status of the member and the reputation of the RICS.

Further, the Committee directed that the attention of members generally should be drawn to the need to be satisfied, when contemplating accepting instructions, that their professional experience was such as to enable them to carry out the instructions competently; and that the time available was sufficient to enable them to carry out the instructions satisfactorily, particularly in instances where the work involved was not within their normal sphere.

The Committee also wished it to be emphasised that in all cases, reports should be carefully checked by a responsible person before despatch."

It must be a matter of concern to all Chartered Surveyors that their Professional Practice Committee felt it necessary to issue such a statement at all. Members of the RICS and other surveyors will appreciate the emphasis thus placed on the maintenance of the highest possible standards of presentation and technical understanding at this point of contact between the profession and the general public.

The Practice Note 'Structural Surveys of Residential Property' makes certain specific recommendations regarding surveyors' reports on residential property, Reports should, *inter alia*:

(1) Be clearly typed with headings and sub-headings to aid digestion and facilitate future reference, with each page indicating the address and reference of the property.
    (I would add to this that pages and paragraphs should be numbered clearly to facilitate reference and avoid the possibility of pages losing their sequence or being mislaid should the report be taken apart for copying. It would clearly be a very serious matter for an incomplete report, or copy, to be relied upon by a client or a third party).
(2) Be prepared in simple, direct, grammatical English free from specialist technical expressions unless they are explained. Be presented in a logical sequence which can be readily followed by the reader, and allow a variation of emphasis to be applied to matters of varying significance.
(3) Be factual and unambiguous wherever possible, distinguishing between fact, the surveyor's firm opinion and any speculative comments included for guidance or discussion.
(4) Should provide a record of the building, its construction and materials. Compare these with modern standards and other properties of a similar type, giving a balanced view of the property which describes the separate elements sufficiently to identify their position, construction and condition.
(5) Should advise on necessary repairs or modifications needed to remedy defects, further investigations which may be required and the future development of defects which could be significant.

Some surveyors' reports include a list of approximate buildiing costs likely to be involved in undertaking specific recommended works. They may, in conclusion, provide a summary of items of disrepair with a list of costs. This approach could sometimes be misleading and place the surveyor in a difficult position if the actual costs which arise differ significantly from his estimates. As a general rule, attempts to cost building works at this initial stage are unwise. There are two reasons for this.

First, in order to price many items with reasonable accuracy the surveyor will need to measure the work involved and prepare approximate quantities which will involve time and will increase the fee. It may be that on receiving the report the client may decide not to proceed with the purchase, in which event the need for such estimates will not arise. It is preferable, in the first instance, to present the facts to the client and to prepare approximate quantities in the form of a supplementary report at a later stage if requested.

Secondly, there are many items of building work which can be undertaken in a variety of ways depending upon the client's requirements, and in the absence of full knowledge of the client's needs it may be impossible in practice to assess the costs accurately. If estimates are provided, the Practice Note recommends that the surveyor should be careful to state at some length the reservations and limitations of such advice.

In some cases, however, a surveyor may be commissioned by his client to specify and supervise any subsequent remedial works which may be found necessary as a result of his survey. In such circumstances, when the surveyor would have control of the works, an estimate of costs at survey stage would be admissible.

In the drafting of a report for a private client, the need for simple, direct, grammatical English, free from specialist technical expressions and capable of being readily followed by the reader is paramount. There will, of course, be occasions when a surveyor will need to prepare a report for a fellow professional in circumstances where technical expressions may rightly be used, but such situations will be unusual for most practices although perhaps more common where pending litigation is involved.

Tenses

Difficulties sometimes arise with the correct use of tenses when drafting reports. Often a mixture of tenses has to be used and there is some difference of opinion as to the best way in which this may be arranged. For example, it is reasonable to say, 'The main roof is of flat construction surfaced in asphalt', or, 'The asphalt surfaces were satisfactory at the time of our inspection', drawing a distinction between the facts ascertained during the survey which apply at the time of despatch of the report and those facts which must be described in the past tense in order to make clear that between the time of the survey and the publishing of the report, events could occur which would invalidate the comments made.

To say 'the asphalt surfaces are satisfactory' may be unwise since this indicates their condition as at the date of the report, correct in most cases but which could be incorrect in unusual circumstances.

To say 'The main roof was of flat construction surfaced in asphalt' is a form used in many surveyors' reports where a practice is made of presenting all facts and description in the past tense, but this can engender a note of uncertainty in the mind of the client who may feel that this suggests that the roof, in some way, has changed. Similarly, 'The main walls were of cavity construction' suggests that perhaps they are no longer so.

Matters of description and descriptions of the construction of a building are best presented in the present tense. Those sections detailing the investigations made and the conclusions drawn should be presented in the past tense as applicable only to circumstances as at the time of the survey, especially in matters which may change significantly in a short space of time such as the state of roof coverings in storms or damp levels in walls following wet weather.

Generally, sentences and paragraphs should be kept short with adequate punctuation to facilitate the reader's understanding. It is no disrespect to the average client to say that he, or she, may well be unfamiliar with the reading of any type of document and that a surveyor's report may seem initially to be a formidable item to digest.

Nomenclature

It is often difficult for a surveyor to appreciate that expressions and terms appearing to him self-explanatory may cause confusion to a client and may require clarification. Some mention may be made by the surveyor in his report of the fact that the client should not feel inhibited from discussing any points needing clarification and that any time spent explaining such matters to clients be regarded as an essential part of the service. Most clients will understand what is meant by brickwork but may not know what 'pointing' is, nor may they understand expressions such as 'plumb' although they will understand what is meant by 'level'. To some extent the degree of technical expression used in the report may be tailored to suit the client if the client's business or profession is known. In one instance, some time was spent explaining to a client in simple terms a minor point of structural design in a building. The client sat patiently through the description only at the end revealing, in passing, that he was a chartered engineer specialising in structures. The structure did not collapse but the surveyor almost did!

The skill needed to draft a well-presented, rounded report is fundamental to the practice of a surveyor who inspects and reports upon buildings. One who has great technical ability and who may undertake inspections and investigations thoroughly may still be unable to present the information needed by his client unless he can produce in the final report a clear, concise and readable document which provides all the information needed.

It is a matter for argument that the preparation of the report is more difficult than the surveying of the building. Both involve degrees of concentration over an extended period if the end result is to be satisfactory. The former is undertaken in the relative comfort of the office whereas the latter is sometimes carried out under inhospitable conditions and may be physically tiring. However, making the inspection may become a matter of procedure and habit whereas the drafting of the report is specific to a particular property and client and that process requires a high level of concentration in circumstances where distractions may arise. The result will be more satisfactory if a report is drafted in a calm, quiet atmosphere free from distraction of any kind – such situations are rarely achieved in a busy surveyor's office.

Drafting the report

Given that the inspection and other investigations have been completed, the surveyor will then have to sit down and commence the drafting process. How should this be carried out?

The surveyor should have notes, photographs, the file and all the relevant information. He or she should have available and close at hand an office library of technical papers, textbooks and trade information. He or she should have the client's instructions and a folder or binder containing all the standard report headings, sub-headings and clauses used by his practice in the preparation of their reports. He or she will have a word processor, a pad of lined paper or an empty spool or cassette of tape in a dictation machine. Some draft their reports on paper but dictation is equally suitable, as is direct drafting from a display screen.

If it were possible to recall the single most lasting impression left by some years of surveying buildings and preparing reports, that chosen by the writer and colleagues with whom this was discussed is that of the empty VDU screen waiting to be filled with facts and information. What a hard taskmaster that VDU screen can be, especially in the early years of his or her professional life when the routine for such matters has not yet evolved and become set.

On such occasions, standard headings, sub-headings and clauses are referred to. Hopefully these will have been carefully updated and improved over the years to provide a useful check-list of matters which need to be covered. The order of presentation is then established together with standard caveats and exclusion clauses. Soon the thoughts will begin to flow and a report draft begins to emerge from the gloom of a mental block. The first rule of drafting might, therefore, be described as having a clear framework within which to work.

In the section of the report dealing with the condition of the building, there should be an order of presentation consisting of:

(1) the investigations made,

(2) the facts ascertained from those investigations, and

(3) the advice which follows from the facts.

For example, 'We carried out a careful examination of the wall surfaces inside and out' conveys to the reader the exact nature of the part of the building investigated. The reader will appreciate immediately that material beneath the wall surfaces will not have been inspected and that the advice given will relate to the information obtained from an inspection of the wall surfaces only.

Then, 'A pattern of small fracture lines was found to the left side elevation, generally following the mortar joints in the brickwork but with two fractured bricks to the main central crack line. Cracks were between hairline and 2 mm wide in three main lines of separation indicating a total extent of separation in the brickwork of about 5 mm at the eaves, tapering down to nil at the damp-proof course line'. Thus the reader will follow the broad gist of the factual information obtained. 'No evidence of cracking to the other elevations is apparent nor is there any such evidence inside the house'. 'An examination of the left side elevation with a level and plumb line did not indicate any evidence of undue separation or distortion of the structure'.

Finally, 'Having regard to the fact that the building is new, it is not possible for us to confirm whether the tensions in the left side elevation are due to initial consolidation or thermal movement in the new work, or whether they indicate some defect below ground. At present the degree of separation in the wall is not severe and if it were found in an older structure which had been on the site for many years it would not of itself be significant. Having regard to the fact that the structure is new, we would recommend that we should excavate a small inspection pit at the side of the left elevation to confirm that subsoil conditions and foundation design are as described in the plans and calculations provided for our use. We also recommend that the left side elevation be subjected to a period of observation to confirm whether or not movements in the brickwork have ceased. The extent and cost of such further investigations are beyond that of this present report and we must await your instructions in this respect'.

In the foregoing examples the recommended order has been followed so that the manner of arriving at the advice may be appreciated by the client involving the maximum economy of words. A client receiving the advice may be disappointed that the result is not clearcut but at the same time he should be under no illusions as to what is being recommended. If he fails to take the advice, the surveyor will not be responsible if, at a later stage, some fundamental defect becomes apparent under the left elevation.

The second rule of drafting might therefore be to describe the investigations, indicate the facts obtained and give advice related to those facts (including advice to undertake further investigations if appropriate).

A statement such as, 'The main roof is in fair condition at the present time for its age' is much less satisfactory than, 'We inspected within the accessible roof void using an inspection lamp and we examined the roof slopes externally from an extension ladder. The roof covering was weathertight at the time of our inspection and the tiles were in satisfactory condition for their age. We would not anticipate that any general re-covering of the roof will become necessary in the foreseeable future but the occasional renewal of tiles may be needed and a periodic check of the roof slopes would be recommended using a ladder, say each autumn when the gutters are cleared out'. This gives a client the correct impression of thoroughness and soundly-based advice.

An alternative method of carrying out a structural survey and preparing a report is to dictate the report direct on to a portable tape recorder while carrying out the survey. This method is not recommended (see Watts and Another v. Morrow in Chapter 19).

There are two reasons why direct on site report dictation can cause problems for the surveyor unfortunate enough to find himself or herself subject to a claim.

Firstly there are no notes on file. The report is all there is. There is nothing to record what the surveyor did or did not do other than what appears on the face of the report. In the event of a negligence claim it may be difficult for the surveyor to defend the report if it appears that something has been overlooked. Notes taken using a suitable field sheet and checklist will at least demonstrate methodology and confirm that all relevant matters have been considered.

Secondly there is no time for what the judge in Watts v. Morrow described as 'reflective thought'.

For example the surveyor may note a damp patch and a test using a moisture meter may show a high reading. Since damp can be caused by moisture rising, penetrating or descending and might result from problems with the roof, gutters, pointing, damp proof courses, bridged cavities, condensation or plumbing, a number of other areas of inspection have to be drawn together. If each aspect is described in isolation when it is inspected without bringing the advice together in a coherent format the end result could be an essentially descriptive but disjointed report which leaves the client none the wiser as to what may be important and what needs to be done.

The writer takes part of his notes in the form of sketches. So there may be a simple line drawing of each floor level indicating door and window openings, chimney breasts and other points of interest. Also a simple sketch of each elevation showing the outline, openings and obvious defects such as cracks or bridged damp proof courses.

Back at the office the damp patch can be seen in relation to both plan and elevation. If the property has a solid wall and there is a cracked rainwater pipe collar outside adjacent to the damp patch inside, an appropriate paragraph can be drafted following the recommended rules: first what was seen, then what appears to be the cause and finally some advice as to what needs to be done.

At the back of the surveyor's mind is always the fear that he or she may have missed something of importance – something could have been omitted from the report which would have serious future consequences either for the building, its occupants or for his or her own professonal indemnity policy! As well as cutting down the element of risk by using the survey method described, good note taking is a distinct asset, as was pointed out by the late Mr Gerald Eve. He advised that when inspecting a property, if anything unusual is noted – no matter what it is – it should be written down. For example, in the case of a surveyor carrying out an inspection of a property he says, 'To your surprise, you discover in a third floor attic at the end of a passage a green parrot in a cage. *Write it down*: that parrot may come in useful'. For example, if something goes wrong and evidence is required in Court –

Question (in cross-examination): Did you go all over the house Mr Sparrow?
Answer: Yes.
Question: All over it?
Answer: All over it.
Question: Including the third floor, Mr Sparrow?
Answer: Yes, including the attic at the end of a passage which is the home of a green parrot.

Keen observation and comprehensive notes are indeed useful!

Unnecessary words should be avoided and the message should be kept as plain and simple as possible.

As an example, there is the quoted instance of a ten-year-old child who was invited to write an essay on a bird and a beast:

The bird that I am going to write about is the owl. The owl cannot see at all by day and at night is as blind as a bat.

I do not know much about the owl, so I will go on to the beast which I am going to choose. It is the cow. The cow is a mammal. It has six sides — right, left, front, back and upper and below. At the back it has a tail on which hangs a brush. With this it sends the flies away so they do not fall into the milk. The head is for the purpose of growing horns and so that the mouth can be somewhere. The horns are to butt with, and the mouth is to moo with. Under the cow hangs the milk. It is arranged for milking. When people milk, the milk comes and there is never an end to the supply. How the cow does it I have not yet realised but it makes more and more. The cow has a fine sense of smell; one can smell it far away. This is the reason for the fresh air in the country.

The man cow is called an ox. It is not a mammal. The cow does not eat much, but what it eats it eats twice, so that it gets enough. When it is hungry it moos, and when it says nothing it is because its inside is all full up with grass.

The simplicity of the child of ten years makes his description clear, concise and even somewhat stylish. Contrast this with, for example:

> The attitude of each, that he was not required to inform himself of, and his lack of interest in, the measures taken by the other to carry out the responsibility assigned to such other under the provision of plans then in effect demonstrated on the part of each, lack of appreciation of the responsibilities vested in them, and inherent in their positions.

This, of course, means that neither took any interest in the other's plans, or even found out what they were. This shows that they did not appreciate the responsibilities of their position.

Keep clear, clean-cut and concise.

To sum up, probably the most important part of the building survey is the report which, if not written out correctly or accurately, can cost the surveyor dear in terms of his professional indemnity!

A scrupulous report writer in every sentence that he writes will ask: 'What am I trying to say? What words will express it?'. He will probably ask himself, 'Could I put it more briefly?' You are not obliged, however, to go to all this trouble. You can shirk it by simply throwing open your mind and letting the ready-made phrases come crowding in. They will construct your sentences for you, even think your thoughts for you to a certain extent and, if required, will perform the important service of confirming your meaning to yourself.

A written report should always be made although it is often necessary to make a verbal report to the client immediately after the survey, particularly if the client is under pressure to purchase. However, if such a verbal report is made, the client must be informed that the written report will take precedence over whatever you happen to have said to him over the telephone or in person.

As previously intimated, the report must be written in a precise manner without relying on obscure technical terms and must be couched in language which is understandable to the layman. A client's particular circumstances must also be taken into account when writing such a report. In other words, for the ordinary lay client, a non-technical report must be used as far as possible, whereas if writing a report for an organisation where fellow professionals are involved, many more technical terms may be employed in order to elucidate a situation.

When the description, age and design of the property are given these could be related to the cost of the annual maintenance to such properties.

The construction of the property, together with any defects which exist or are likely to arise, must be fully described and elaboration must be made of any defects so that the probable cause and the necessary remedies are immediately apparent to the client.

The points on which the surveyor is not able to advise must be set out clearly, and positive recommendations made for further inspection and reporting by specialists if this is considered necessary.

Disclaimer and exclusion clauses

Normal practice requires that all surveyor's reports include disclaimers of liability or exclusion clauses and the surveyor's professional indemnity insurers may well insist that any report which deals with the condition of a building contains such clauses as a condition of insurance cover.

The distinction between a disclaimer and an exclusion might best be seen in relation to a dry cleaning shop: if there is a notice on the wall saying 'we do not guarantee to get your clothes clean' this would be an attempt to disclaim liability for poor results; if it says 'we do not clean the right sleeves' this would be an attempt to exclude liability altogether for that part of the garment.

The following are typical examples of standard warnings, disclaimers and exclusion clauses commonly employed at the present time – it should be emphasised however that all such clauses are subject to the tests of reasonableness under the Unfair Contract Terms Act and liability may still arise (see Chapter 19).

Third parties   This report is prepared for the information of the addressee only and without responsibility towards any third party.

Limit of inspection   The inspection did not include any parts of the structure which were covered, unexposed, inaccessible or below ground and we are therefore unable to report that such parts are free from defect.

Subsequent changes or new information   Advice and information given in this report is prepared on the basis of conditions prevailing at the time of inspection including, where applicable, specific weather conditions and the limitations on inspection imposed by fitted floor coverings and other obstructions. No responsibility can be accepted for changes in the condition of the property which may have taken place after our inspection or for matters which may only come to light following exposure of the structure.

Concretes   No special tests have been made at this stage on cements and concretes used in the construction and accordingly we are unable to report that concretes are of suitable strength and free from the presence of High Alumina Cement, Chlorides, Sulphates or other deleterious materials; as regards concretes below ground we cannot confirm that they will be suitable for the ground conditions if the subsoil contains Sulphates or other damaging constituents.

Locations and dimensions   Directions right or left are always taken as if viewed facing the property from the highway at the front. All measurements and dimensions are approximate or nominal only and should not be used if accuracy is required.

Insurance   All property owners are advised to ensure that their property is insured, from the moment of exchange of contracts, for a sufficient sum, against all usual perils including Fire, Impact, Explosion, Storm, Tempest, Flood, Burst Pipes and Tanks, Subsidence, Landslip, Ground Heave and Public Liability.

Clarification   If further clarification of matters raised in this report is required, or additional advice regarding the property, please do not hesitate to let us know.

Fixtures and fittings   If there is any possibility of doubt as to the contents, fixtures or fittings included in the sale we recommend that Vendor and Purchaser agree a list and you should confirm the position regarding light fittings, floor and window coverings etc. You are also advised to carefully examine any fixtures, fittings or contents included to confirm whether these will meet your requirements.

Electricity   Prior to exchange of contracts we recommend that you obtain a report on the installation from a qualified electrician and that any work necessary to comply with current Institute of Electrical Engineers' Regulations be undertaken.

Gas   Prior to exchange of contracts the gas service together with any gas appliances included in the sale should be inspected and tested by a CORGI registered contractor and appliances should be serviced before they are used.

Central heating   Prior to exchange of contracts a report from a qualified heating engineer should be obtained and he should be asked to confirm whether the system will provide heating and hot water to meet your requirements. The advisability of using corrosion-proofing chemicals in the circulation should be confirmed and the suitability of the system for a regular maintenance contract.

Conclusion   We see no reason why you should not buy this property if the price is reasonable and if the various matters raised in this report are appreciated including (where applicable) the limitations of the inspections and tests made at this stage.

Chapter 11

Pro-forma reports

In view of the great variety of building types, designs and construction methods, no simple pro-forma report can convey effective, balanced and complete advice.
RICS Practice Note, 'Structural Surveys of Residential Property'

Can you be sure that the house you are buying is free from major structural defect or that it represents a sound financial investment? The simple answer is that you *can't* be sure. The RICS' 'House Buyer's Report and Valuation' is designed to remove these doubts
Buying a House?, RICS, 1981.

During the course of his or her work the surveyor may be required to complete a wide variety of pro-forma reports. Initially, such reports were used only by the building societies for their own confidential use. More recently, such reports have been introduced by banks, insurance companies and other lenders. In 1981 the provision of pro-forma reports became a service specifically sponsored by the Royal Institution of Chartered Surveyors with the introduction of their 'House Buyer's Report and Valuation'.

In that same year the major building societies also adopted a new practice of sending a copy of their own surveyor's report to the mortgagor for his, or her, own information, subject to various caveats and warnings concerning the limitations of the mortgagee's surveyor's inspection.

These changes came about as a result of pressure from house buyers for simple straightforward advice on the property in which they were interested. It was found that only about 10% of house purchasers took independent professional advice before buying their houses. Most took no advice at all but relied upon the building society to tell them if anything major was amiss; alternatively, they simply relied upon their own judgment.

Building society, insurance company and bank reports are, in most cases, intended to be used for all types of residential and small commercial properties. The space allowed for answers is often insufficient so that in completing the report, the surveyor will often find it necessary to insert 'See separate sheet' and then provide an addendum to the pro-forma report to include a specific comment or describe a defect. This appears to be universal practice among surveyors with a regular flow of building society and other pro-forma report work.

A similar situation arises with the RICS' 'House Buyer's Report and Valuation' which may be completed easily in those cases where the house is sound and in good order but which requires careful use where some significant defect is found or suspected. In such circumstances, continuation sheets are provided to enable extended comments to be added where these are necessary.

The RICS scheme is for an inspection of the property broadly the same as the basic inspection undertaken for a full building survey report. It is unlikely that this inspection of itself will take less time than a normal full inspection. The RICS scheme is intended for houses and bungalows up to 200 m² in area and not more than three storeys high. The scheme is not intended for period properties. The RICS scheme includes a set of pre-printed working report forms on which the surveyor writes his notes during the actual inspection, together with triplicated, 6 or 7 page report forms and folders for the clients.

There are certain specific limitations on the inspection included in the RICS' 'House Buyer's Report and Valuation' scheme. The inspection will not include underfloor areas or a close examination of roofs more than 3 m above ground level. No tests of services are included. The provision of the service is subject to certain specific conditions of engagement set out in an explanatory booklet, *Buying a House?* published by Surveyors' Technical Services.

On the following pages will be found a fairly typical example of a pro-forma report form which a surveyor might have to use when reporting for a prospective mortgagee. Since there is an almost infinite variety of report forms used in practice, it will be found that the questions asked vary considerably from case to case. However, there are certain questions which are common and which will therefore be generally encountered. These are:

(a) description of the building and its locality, including age,
(b) detailing the accommodation and services provided,
(c) requesting information on the construction and condition,
(d) requiring dimensions or a plan for identification,
(e) requiring a valuation for mortgage purposes, and
(f) requiring a valuation for insurance purposes.

As regards questions aimed at obtaining a description of the building and locality, the reader will not normally have seen, or have intended to see, the property concerned and any comments, favourable or adverse, likely to assist the reader in forming a mental picture of the property would be helpful. The general age and character of the surrounding neighbourhood is crucial to the value of most types of property and is significant in considering whether or not the future trends in values will be satisfactory. If a house is modern but surrounded by much older and more run-down houses, this could be important. Such considerations will affect not only the valuation for mortgage purposes but also the percentage advance likely to be made and the term over which that advance may be recommended.

Measurement of the accommodation is often required, although generally given approximately on estate agents' particulars. The surveyor will then have to measure the rooms, a task which is not always as simple as it sounds since special treatment is needed for bays, alcoves, projecting fitted wardrobes and the like in order to convey accurately the room size and provide a fair comparison of room sizes between different properties where records are kept for valuation or statistical purposes. For such measurements in all types of property the Royal Institution of Chartered Surveyors and the Incorporated Society of Valuers & Auctioneers have jointly published a 'Code of Measuring Practice' and it is recommended that this Code be observed where possible.

The writer would normally measure rooms to include any fitted wardrobes and also chimney breasts but exclude bay windows and other small projections. Such dimensions have an advantage of being more likely to reflect the floor spans above as well as the room dimensions should a cross-check on floor spans be needed.

Concerning information on construction and condition, most of the questions and answers will be self-evident but a distinction needs to be drawn between minor defects which are commented upon, defects which might be subject to some condition on the advance, defects which are sufficiently serious to warrant a retention from the advance pending their remedy, and defects requiring further investigation before an advance can be confirmed.

The purpose of requiring a plan or the main plot dimensions is to confirm that the title plan corresponds with the property inspected. On occasion in the past attempts have been made, by fraud, to obtain a mortgage advance on the security of a property title and to encourage a surveyor to inspect quite a different property, perhaps by changing over house numbers or names. Occasionally it will be found that a house with a large garden or paddock will be inspected and that some, or most, of the grounds will be excluded from the title, being separately owned and rented under a tenancy or licence from an adjoining property owner. Obviously, if the property is mistakenly valued on the assumption that the garden or grounds are substantially larger than they really are, then this could be a very serious matter indeed.

A valuation for mortgage purposes should have regard to all matters ascertained by the surveyor during his inspection, including the condition of the property, and may be expressed as a 'forced sale' valuation, a 'current open-market value' figure or in some other way, depending upon the lending practice of the prospective mortgagee. Such a valuation may, or may not, be accompanied by an opinion from the surveyor on the recommended term over which a mortgage may be granted and the percentage level of advance appropriate, having regard to all the circumstances.

A valuation for insurance purposes will normally be required on a reinstatement basis, assuming a total loss and rebuilding of a comparable structure of similar type and appearance complying with the current Building Regulations. This figure will not have any direct relation to a valuation for mortgage purposes. It may be lower than the mortgage valuation if the building is modern and in an area of high land values. Frequently it will be higher, sometimes substantially higher, than the market value because reinstatement would require expenditure on a substantial scale in order to reproduce a building with high headrooms and other features no longer considered economic. Compliance with current Building Regulations or London Building Acts will have to be assumed and planning requirements taken into consideration. The whole question of valuations for insurance purposes could take chapters to discuss adequately and is beyond the scope of this volume but these are generally assessed on a rate of £/m^2 varying with type, age, use and situation of the property.

One further point regarding pro-forma reports for prospective mortgagees is that such reports may occasionally have to be issued in a provisional form leaving certain questions unanswered pending the results of further enquiries or investigations. Occasions when this may arise might be:

(a) If a structural fault is found or suspected, this requiring investigation and specialist reporting. In such a situation, a mortgage advance may be recommended subject to the mortgagor obtaining the necessary further advice or paying for the commissioning of a report from a building surveyor, consulting engineer or obtaining a laboratory test report. If the surveyor is not satisfied with the results of his own inspection and if he is required to return his report promptly, there would appear to be no reason at all why a provisional report conditional upon such further investigations or tests may not be immediately issued.

(b) If the mortgage security is affected by some other potential defect such as a possible planning scheme, road widening or other constraints which could affect the value, a provisional report might be issued subject to further investigations by solicitors being completed.

On the following pages we reproduce a typical example of a pro-forma Residential Property Mortgage Valuation Report published by agreement with The Nationwide Building Society. This particular form was introduced by the Society in 1988 for residential valuations.

In keeping with the generally accepted practice of major United Kingdom banks and building societies, the form is issued in triplicate with one copy for the society as mortgagee, one copy for the borrower as mortgagor and one copy for the surveyor for retention on his files.

It will be noted that there are important notes for applicants (the borrowers) in conditions which they sign, these being important exclusion clauses intended to limit the liability of the surveyor to the borrower. These clauses are very clear and unambiguous; there should be no doubt in the mind of the borrower as to the nature of the inspection or the nature of the report. Nevertheless, these clauses would be subject to a test of reasonableness under the Unfair Contract Terms Act and the judgment of Mr Justice Park in *Yianni* v *Edwin Evans & Sons* (see Chapter 19).

It will be seen that the valuer must state that he or she is not contravening Section 13 of the 1986 Building Societies Act, which means that he or she has no interest directly or indirectly in the property being offered as security. Inspections for building societies and other mortgage lenders are normally subject to specific conditions of engagement agreed between the surveyor and the lender. Also the Royal Institution of Chartered Surveyors and the Incorporated Society of Valuers and Auctioneers have jointly published *Mortgage Valuation Guidance for Valuers* to apply to inspections carried out for mortgage lenders.

Clearly it is essential that a surveyor undertaking this work should have no connection whatsoever with the vendor or the purchaser so that advice given may be seen by all parties to be dispassionate. In the event of a surveyor's advice being called into question and a relationship found to have existed, this could seriously undermine the ability to defend his position, both in connection with a mortgage valuation or any other type of report.

In the past, pro-forma reports were prepared almost exclusively for mortgagees on a completely confidential basis with no liability assumed towards third parties. This situation has now changed for these three reasons:

(a) The introduction of a service of providing pro-forma reports for prospective purchasers,

(b) the policy of the major lending institutions of releasing a copy of their reports to the borrowers, and

(c) the legal developments in recent case law which have called into question the extent to which a surveyor to a prospective mortgagee may limit his liability towards the prospective mortgagor. (See Chapter 19).

Whatever the future may hold in the form of new legal precedents or legislation, it is clear that the traditional confidential surveyor's report prepared without responsibility towards any third party is no longer viable. There are reasons to regret this, but the reasons why this situation has come about are entirely logical. Not the least is the fact that most lending institutions were in the practice of collecting a survey fee from the borrower with the application for the loan and borrowers felt aggrieved that they were not permitted to see the reports for which they themselves had paid.

Certain basic questions raised by these changes in climate are:

(a) Who should be regarded as qualified to prepare pro-forma reports?
(b) What minimum standard of inspection should be considered necessary?
(c) How much detail should be included in the completed report?

In the case of the RICS' 'House Buyer's Report and Valuation' this may only be prepared by a chartered surveyor, and the qualifications and experience of such a surveyor should be the same as those which are required for building survey reports of the traditional type. Other professional bodies in this field also promote a service for their members along similar lines.

The standard of inspection for the RICS scheme will generally be the same as for a normal building survey but subject to certain specific limitations laid down in the conditions of engagement. In *Cross and another* v. *David Martin & Mortimer* [1989] IEGLR 154 Mr Justice Phillips, sitting in the Queen's Bench Division of the High Court, had to consider a case concerning a House Buyer's Report.

The plaintiffs had alleged that the surveyors had negligently failed to draw attention in their report to three features of the property, namely settlement of solid ground floor slabs, misalignment of a number of first floor door openings and alterations which had weakened roof trusses when the loft was converted into a room. The surveyors did not dispute the existence of these defects but contended that the first feature was not ascertainable during a House Buyer inspection (which excludes lifting fitted carpets) and the other two features were not sufficiently serious to require comment in the report.

The surveyors lost the case. It is unwise to attempt to read too much into one case based upon its own set of facts, nevertheless the outcome appears to confirm that the courts would view a House Buyer's report as a Building Survey presented in an abbreviated form rather than a simple valuation with limited added comments on condition. As such it may not be enough to comment on obvious defects which might affect value. It is necessary to include comment on surrounding circumstances which could lead to defects and on defects which the surveyor may not consider serious. In view of this judgement continuation sheets may be increasingly necessary for all but the simplest properties if an adequate report is to be provided.

It would be inappropriate now, as in the past, to assume that unsupervised building society and other institutional inspections may be taken by the less experienced or younger staff in a surveying practice who would not be considered competent to undertake building surveys. The highest standards of professional

| Report Number: | | Instructing Office: | |
| Voucher Number: | | Reporting Office: | |

Residential Property Mortgage Valuation Report
for Nationwide Building Society

Nationwide

Inspection date:

NATIONWIDE SURVEYOR'S COPY

1. Mortgage Details
  a) Property address

    Post Code
  b) Applicant(s)

  c) Purchase Price   £
  d) Advance   £
     Term   years

2. Tenure
  a) Freehold/Feuhold
  b) Leasehold unexpired term   years
  c) Rent i) Ground/chief/feu £ p.a.
      ii) Fixed/escalating etc.
  d) Maintenance charge £ p.a.
  e) Other

3. Tenancies (if any): Give details and rent(s)

4. Amenities (Situation/Locality/Facilities)

5. Description of Property
  a) Detached, semi-detached, terraced
  b) House, flat, bungalow etc.
  c) Year built (approximate)
  d) Number of storeys

6. Accommodation (Listed by floors, include external, garage space, outbuildings etc.)

7. Construction
  a) Main Building
    i) Walls
    ii) Roofs
    iii) Floors
  b) Garage(s)
  c) Other Buildings

8. Services (Including drainage)

9. Central Heating (Type, Fuel, etc.)

10. If New Property:
  a) Name of Builder
  b) NHBC or Other
  c) Stage of construction
  d) Is reinspection necessary?

11. Roads/Footpaths (including side and/or rear)
  a) State if made/partly made/unmade
  b) Retention recommended £

12. Other Matters That May Materially Affect Value
  a) Is the property readily resaleable at or about the valuation figure?
  b) In cases of flats etc. is proper management/maintenance apparent?
  c) Has the property ever been affected by structural movement caused by subsidence, settlement, landslip or heave?
  d) Is the risk of further movement one the Society can accept? (If "No" decline the property)
  e) Any other important factors?

(If "Yes" list below):-

13. Building Insurance
  a) Estimated current reinstatement cost including site clearance and professional fees excluding VAT except on fees £
  b) Also for flats-estimated reinstatement cost of whole block (except Scotland) £
  c) Floor areas-state as external Metric or Imperial
    i) Main Building
    ii) Garage (if not integral)
    iii) Other Buildings
  d) Does the area warrant special consideration in respect of any insurance risk? (eg. contaminated land, flood etc.)
  e) Number of "habitable" rooms (as guidance note definition)
  f) Detached/Other for insurance rating purposes (see guidance note definitions)
  g) Is this property which needs to be referred due to special insurance risks?

Page 2 of 4

132

Security Address

14. **Matters to be Checked by Conveyancers**
    (amplify if necessary in 15)
    a) Rights of Way/Easements/Servitudes/Wayleaves (where apparent on inspection)
    b) Road Agreements/Liability
    c) In the case of flats etc., a properly formed Management Company
    d) Drains/sewers liability
    e) Other

15. **General Remarks:** (Including condition)

16. a) Works to be Carried Out as condition of mortgage subject to retention below (list)
(only include work absolutely necessary to protect the Society's security. The amount of advance must be ignored)

b) Amount of Recommended Retention (This is not an estimate of costs.
The Borrower(s) should obtain detailed estimates before proceeding with purchase)   £

17. Valuation for Mortgage Purposes - (assuming vacant possession unless otherwise stated)
a) Is the property a suitable security for the Society? (answer Yes or No)
b) Valuation in present condition (do not value if answer to 17a) is "No")   £
c) Valuation upon completion of any works required under 10, 11 or 16   £

18. Other Reports
Have you or your firm prepared any of the following on the property:

a) a NBS Homebuyer Report? (Yes or No)

b) a RICS/ISVA House/Flat Buyers Report? (Yes or No)

c) a Structural/Building Survey Report? (Yes or No)

If the answer to b or c is "yes", indicate whether "before" or "with" this Mortgage
Valuation Report?

I certify that I have personally inspected the property and that in making this report I am not contravening Section 13 of the Building Societies Act 1986

Valuer's Signature:
Name and Qualification:
Firm:
Address:

Post Code:
Firm's Identity Code:
Telephone:
Fax:
Date:

Page 4 of 4
Nationwide Building Society

competence are required for all reports, whether pro-forma reports or otherwise, and great care has to be exercised in all surveying practices to ensure that the routine pro-forma report work reaches a suitable standard.

Although the fees for a pro-forma report may be lower than for other work, this will not absolve a surveying practice from liability. If there is any question that such work be uneconomic then it is suggested that, in view of the current legal position it is better to decline instructions rather than to attempt an over-hasty inspection. It is to be hoped, however, that most practitioners will not decline instructions on these grounds since the hallmark of any learned profession is a willingness to provide a service to clients not necessarily concentrating solely on those activities which are most profitable. In this sense most professionals have to take some of the rough with the smooth.

The question as to the amount of detail which should be included in a pro-forma report is difficult to answer since the intention of using a pro-forma report system is to achieve economy for a client, bank manager or building society manager who will not wish to plough through many pages of descriptive text but wishes to have his attention drawn to the salient matters.

The author recalls the time when as a student he was accompanying an experienced surveyor accustomed to undertaking a large number of building society inspections and who was a master of the brief but salient comment. The main survey having been completed, this surveyor proceeded into the garden of the house to find a boundary wall about a metre high and ten metres long leaning severely towards the adjoining curtilage. The wall was built in 112-mm half-brick thickness with insufficient piers, using a black ash mortar and was clearly on the point of collapse. The surveyor carefully placed his foot against the wall to see if he could induce any movement in it and pressed slightly. The whole wall then proceeded to collapse on to the flower beds in the adjoining garden.

The surveyor was not ruffled by the course of events. He carefully looked about him to see whether or not the incident had been observed (which fortunately it had not) and wrote 'Rebuild west side garden wall' on his pad, before beating a rapid retreat to his car.

By way of contrast on another occasion the inspection was being made with a surveyor who liked to elaborate a little and the inspection was being made of an old and dilapidated cottage in an isolated country location. The report was for a bank who were going to lend on the basis of the borrower's proposals for restoration and renovation of the property. In his report the surveyor began, 'I can, perhaps, convey the flavour of the property best by saying that it was not necessary for me to borrow a key in order to inspect it because there are, at present, no doors'.

Chapter 12

Reports on non-residential buildings

We're taking a full repairing and insuring lease, but I'm sure that's all right. They are in rather a hurry so I have agreed to sign the contract this afternoon. I thought that you could just give it the once-over and 'phone me back before lunch. Incidentally, he says that the flat roof has been leaking but it's all right now.
Potential client's telephone instruction

The basic principles described for surveying and reporting upon dwelling-houses also apply to other types of building, the main distinctions being those of size and usage, although a larger industrial building may, if modern with open headroom and clear spans, be easier to inspect than a small house.

The larger older buildings will often require a lengthy inspection, perhaps lasting several days and the surveyor will need to ensure before undertaking such work that he or she can reserve sufficient time in the diary to complete the inspection to the necessary standard.

A special consideration arising in the case of the larger commercial buildings is the importance of concealed structural elements. Before a client can be advised on the construction of such a building and before any structural alterations can be contemplated, it is essential to determine the structural strength of the building, including the manner in which the loads are carried and transferred to the subsoil.

Where concealed structural elements are involved, the surveyor will have to undertake an initial inspection and then he may have to proceed with a process of research and detection in order to discover the facts. Such research will normally include searching for any existing architects' or engineers' drawings (in one case a most useful set of such drawings were found in a wall cavity), visiting the local authority offices or district surveyor to see if any drawings are available for inspection, and even perhaps opening up the building in question to confirm the position.

If the building has an engraved plaque or stone with the name of the original architect indicated, then the first step would be to see if he or his firm can be traced. The present occupiers or owners of the building may have been there since completion in which event they should know the identity of the original designer and perhaps even own a set of plans. Even if not the original owners, it is possible that plans, or information on the location of plans, may have been passed on to them by their predecessors.

If the present owners are a large company, contact with a long-serving member of the staff may help since such an individual may provide useful information if he can recall the building during construction. For example, the names of builders, architects, manufacturers of fittings, sub-contractors and other vital information.

Working drawings for steelwork may be obtained if the identity of the original steelwork sub-contractor can be discovered, since such contractors often keep records going back for many years. A large number of steel-framed buildings were erected between 1925 and 1949 where such a sub-contractor, if located, may be the only likely source of information.

Having exhausted these sources of original drawings, the surveyor could then go on to see if information is available from the local authority. Local authorities are now much more reluctant than in the past to permit an examination of their records. Such information as is available is often microfilmed and during that process, many older plans and documents may have been destroyed.

From the local authority point of view, there are two problems affecting a general inspection of their files and records.

First, it is considered that in order to avoid a breach of copyright no copying of deposited plans of any kind should be permitted without the original owner's or architect's consent. In order to take sketches from a deposited plan one has, on occasion, been required first to obtain a signed authority from the original author of the drawing. In many cases obtaining any such authority is impossible, particularly in the case of older buildings.

Secondly, local authorities are becoming increasingly subject to litigation over building defects and will often fear that attempts to examine closely approved plans may result in legal actions should any errors or omissions be found. A general reluctance to permit any inspections at all has, as a consequence, arisen with local authorities in some areas.

A surveyor who is a member of the local authority staff will therefore be in an advantageous position in this respect when investigating and reporting for his authority upon buildings in his area.

Having thus obtained all available information in the way of plans and working drawings and having carried out an initial survey of the building concerned, the surveyor will then have to consider the extent to which the building ought to be opened up in order to complete his report.

By this stage, it will have been possible for the surveyor to have formed a view of the superficial condition of the building and services, also the extent to which it complies with current standards for means of escape in case of fire, fire precautions, the Factories Acts, the Offices, Shops and Railway Premises Act and other legislation so that it may be that a preliminary report could be prepared for the client on this basis. If so, it is essential that the client be warned that no opening up of the structure has been undertaken and that the condition of the main structural elements has not been confirmed.

Post-War office blocks and other structures constructed in reinforced concrete present a particular problem and considerable care needs to be exercised if the structure has to be disturbed. Reinforced columns and beams cannot be opened up easily or exposed to see how reinforcement is arranged and such reinforcement is essential to the strength of the structure. If pre-stressed beams are suspected, it is most necessary that these should not be cut into or otherwise disturbed. In such circumstances, a surveyor should measure the overall sizes and spacings of the columns, the dimensions and spans of the beams and then undertake a surface inspection for evidence of chemical deterioration and like problems. Beyond this he would be bound to refer his client to a specialist consulting engineer so that any core samples may be taken or special tests made to detect the presence of high-alumina cement, chlorides, sulphate attack or other such faults having regard to the known facts.

Steel-framed office buildings and other structures dating from the inter-War and post-War years may be opened up more easily than reinforced concrete structures, and the quality of the original construction and the condition of the steelwork may then be judged. It is arguable as to how far one should take such investigations but if the surveyor can confirm sizes, spacings and spans of principle members, together with the loads imposed, it should be possible for him to obtain a check by calculation from a structural engineer and thus reassure his client that the steel-framing is technically adequate and, where seen, in satisfactory condition and protected against corrosion.

It is clearly essential for the client to be advised whether or not a commercial or industrial building will satisfy the Fire Precautions Act 1971 and the Fire Precautions Act 1971 (Modifications) Regulations 1976, together with any subsequent or additional legislation having regard not only to the results of the building inspection but also to the specific needs of the particular client and the use he proposes to make of that building. The provision and maintenance of fire precautions and means of escape form an increasingly important part in the day-to-day management of commercial and industrial buildings. If a need for certification under the Act arises, whether from a proposed change of use, structural alteration or other reason, the client will need to know in advance whether or not difficulties in obtaining certification could arise and exceptional costs occur in adapting the building to meet certification standards.

It is necessary to consider whether or not exceptional energy costs are involved in lighting and heating the structure under review, and the surveyor should then inspect the existing space heating appliances. Generally, a specialist report from a heating engineer should be recommended and some investigation then carried out into past energy costs. In most cases, the vendor or existing lessee of the building will be able to provide details of accounts for oil, gas, electricity and other energy inputs into the building and these may be used as guidance to current cost levels as the building is presently used. Care should be exercised when assuming that current energy cost levels will apply to that building in the future, and if the client's needs change then a change in energy costs may also arise.

At this point the surveyor may have to advise his client on the overall condition of energy-intensive services and whether or not they are efficient, or if the type of energy in use might be changed for a more economic type (for example, gas space heating instead of oil, assuming this can be made available). One should not assume that current relative differentials in energy costs will always be maintained. Energy which is proportionately cheap now need not always remain so and the capital cost of a change-over may not be economic if the differential in cost becomes less advantageous in the future.

As far as energy costs are concerned, the overall thermal efficiency of a building is most significant and should be considered by the surveyor. An industrial structure with a high eaves level, ideal for loading bays and overhead gantries, will be very wasteful of heat as compared with a similar floor space with low headroom. If there is no need for a high ceiling, then the provision of a suspended ceiling could be recommended to reduce space heating costs. This particularly applicable to a single-storey structure or the top floor of a multi-storey structure when, in addition, insulation over the suspended ceiling may be suggested.

When considering the services, thought should be given to the client's proposed use of the building and the numbers and sexes of staff likely to be employed therein. Recommendations may then need to be made over the suitable provision of lavatory and washing facilities. The Offices, Shops and Railway Premises Act 1963 demands sufficient and suitable lighting, a reasonable temperature in every room (15°C), sufficient washing facilities, a supply of hot and cold water, and adequate ventilation. It also covers a variety of other environmental needs. In so far as the premises inspected may fall short of the standards required, the necessary comments and recommendations should then be made.

Older commercial and industrial buildings are generally found to have been constructed in a great variety of ways. There is often a mixture of cast-iron, wrought-iron or steel columns and beams forming the basic structure, and load-bearing brick or stone walls to the exterior and around stairwells and shafts, with timber or concrete floors of various types spanning between the main elements. The concrete floors may be constructed using obsolete methods, often in the form of *in situ* poured concrete construction or part pre-cast and part *in situ* construction. Research into the various forms of construction which are found is a most interesting area of professional activity, particularly in view of the great variety of construction methods likely to be found.

Floors in older commercial and industrial buildings are often of filler-joist construction with a layout of steel or iron beams and an infilling of *in situ* concrete, the beams being either of solid or open-webbed types. In those structures erected before the advent of modern Portland cement concretes, it is likely that a lime mortar could have been used to bind together a clinker, brick or stone aggregate. Some caution is required in assessing the strength of these older types of lime mortar solid slabs since they will be most unlikely to have a strength comparable to the modern Portland cement concrete slabs. This may be significant if an older structure is to be used for different or heavier machinery.

Opening up older structures may be necessary to determine the sizes of the beams and columns, the thicknesses and strengths of load-bearing walls and the nature of the floor construction, including the effective spans of the flexural members and their manner of bearing. Finally, having traced the main forces and loads through the structure down to ground level, a final confirmation of the size, type and adequacy of foundations used may be needed having regard to both present loadings and any different future loadings which may apply. Samples would then be taken of concrete, bricks, mortar, steel, wrought iron, cast iron and timber to confirm the condition of such elements, and these can be sent to specialist laboratories for analysis with a view to confirming the strength of the structure.

In addition to confirming the construction of the building and the nature of the materials used, it is also necessary to confirm the condition of framing since this will be relevant to the strength. Obviously, if an engineer is to be asked to check the suitability of a building by calculation on the basis of the information given as to the loads, spans and materials used, any such calculations will be valid only if the main structural elements are in good condition. Samples taken may or may not be representative of the structure as a whole.

In many old buildings, ironwork or steelwork is buried in the walls in a manner which provides only limited protection from the elements and corrosion may be severe, especially on the weather sides or at points where leaking gutters or other sources of dampness have subjected the walls to constant saturation over many years. The older concretes, consisting of breeze and lime mortars, contain a high sulphate content which will attack adjoining ferrous metals. Roof leaks may cause deterioration to steel roof joists and the upper parts of a steel- or iron-framed structure. Steel columns set into basement construction, or steel feet set into foundations or footings, may be subject to considerable corrosion over the years from dampness inherent in such areas where tanking may have been omitted or become porous. Silver sand for rendering or in a concrete mix and containing a high proportion of china clay fragments and mica is particularly suspect. The effect of weathering is to wash out the china clay content causing the mica to flake away, and the surface of the render or concrete then becoming very porous causes the reinforcement under to rust and crack or even throw off the cover. Readers are recommended to study *BRE Digests* 263, 264 and 265 which deal with the protection, diagnosis, and assessment of deterioration and repair of reinforcing steel in concrete.

In connection with concrete construction there have in recent times been a number of reported failures of concrete elements due to chemical attack or chemical deterioration. In April 1975 the Building Research Establishment published a current paper titled 'High-alumina Cement Concrete in Buildings' (CP 34/75), in response to the collapse of two roof beams in a school in Stepney in February 1974.

The results of investigations published in the 1975 paper indicated that most high-alumina cement (HAC) concrete pre-stressed beams used in buildings more than a few years old is highly converted and chemically changed or likely to become so. The strength of highly converted concrete is substantially reduced.

A considerable margin of safety is, however, provided for in most pre-cast concrete beamwork and slabwork. Proof loading tests in the field and experience in practice resulted in the BRE proposing that for all practical purposes the risk of failure of pre-cast beams or slabs using HAC concrete was very small for spans of up to 5 m.

Accordingly, if HAC concrete is suspected to have been used in a building with spans of 5 m or more, investigation should be made to:

(a) Confirm if in fact HAC was used, and
(b) determine whether the loss of strength due to chemical conversion may be significant for the strength of the structure.

Such an investigation could be undertaken using a simple pull-out test to assess the strength of *in situ* concrete, as suggested in a Building Research Establishment Current Paper published in June 1977 (CP 25/77).

There are several other ways in which concretes can suffer chemical attack or deterioration which could necessitate advising special tests. A concrete mix can also be less than ideal, or variable, and thereby causing a loss of strength. Good building practice is not always observed on site, and concrete foundations laid in wet trenches or mixes which are incorrectly or unevenly proportioned can frequently be traced.

In the event of soil conditions indicating a danger of sulphate attack, special sulphate-resistant concretes should have been used but this is not always so and is generally unlikely in the foundations to older buildings. Deterioration is also possible if the concrete is in contact with other material which contains a high sulphate content, such as certain types of clinker breeze. Chemical attack from chlorides is also possible and occasionally encountered during a survey.

In addition to the possibility of chemical attack or chemical deterioration of the concrete itself, a possibility also arises of corrosion in the reinforcement. Both types of deterioration will be more likely and rapid in damp conditions, often being found to occur together for similar reasons. The amount of protection given to reinforcement in the past was frequently insufficient. Once the concrete cover protection given to the reinforcement begins to crack and soften, then wind and weather can begin to attack the reinforcement itself. Swimming pool roofs are a known problem area as they often have large spans of pre-cast concrete construction situated in a warm, humid atmosphere, and subject to high levels of condensation and chemical attack from chlorine. They may also become subject to weather penetration at defective jointing or on flat roof surfaces.

Before 1977 some concrete work incorporated calcium chloride as an additive to speed up the hardening process. This material absorbs water and forms a corrosive agent which rusts reinforcement bars and forces off the encasement; the combination of spalling encasement and rusting reinforcement may be highly damaging. Testing similar to that advised for HAC is required to confirm the presence or otherwise of this additive.

Occasionally it may be found that woodwool slabs have been used as permanent formwork to reinforced concrete, and this is a potential threat to the strength of the concrete because alkalis and sulphates which may be leached out of the slabs can react with the concrete; such reactions may be especially severe if HAC is present and in damp conditions. In addition, any permanent formwork may hide voids where the concrete has failed adequately to encase the reinforcement because of insufficient vibration during casting.

A significant proportion of commercial and institutional buildings erected since 1950 have been constructed with curtain walling using large glazed areas and spandrel panels; thermal and sound insulation is often poor and the glazed areas often face afternoon sun and prevailing weather so that deterioration to exterior joinery is severe. A large part of Britain's £20-billion stock of school buildings were erected using this type of construction and decaying timber joinery and plywood cladding used in such buildings has been reported as a major problem by 20 organisations surveyed by the Department of Education. Poorly selected softwood was often used, with a high proportion of sapwood and rarely tenoned, dowelled or glued to restrain shrinkage; maintenance has often been poor. In some cases complete replacement of a facade will provide the best repair and offers an opportunity to improve thermal insulation and use better materials.

After reading the foregoing summary the reader may feel that reports on non-residential buildings present rather a formidable challenge to any surveyor's technical knowledge and to an extent this is true – certainly some additional research is required in many cases. The main points to bear in mind are the specific needs of the client and the purposes to which the building will be used; surveyors should always have at the backs of their minds the thought 'What are the possible pitfalls for this specific client in buying or renting this structure? Will he have allowed in his cash-flow for the maintenance and other costs which will arise? What should he be warned about?'

Chapter 13

Reports on flats and other dwellings in multiple occupation

We hear every sound they make in the flat above; and some sounds they don't make as well.
Ground floor tenant

Frequently a surveyor's report is required on a flat, apartment, maisonette or other unit of accommodation which is part of a building divided laterally as well as vertically. Such property will normally be leasehold, this factor alone requiring special treatment (see Chapter 16). Very occasionally a 'freehold' flat or apartment will be encountered in England and Wales but the equivalent of a 'freehold' will be the rule rather than the exception in Scotland where different law applies.

If a freehold flat or apartment in England or Wales is to be surveyed, the client should be advised to discuss the implications of tenure with his solicitor. There are certain technical objections to 'flying freeholds' under English law as it now stands, because of the difficulty of enforcing positive covenants against the subsequent purchaser of a flying freehold who has not himself had any direct contractual relationship with the adjoining owners. In certain circumstances under English law, action to enforce covenants of repair and support can be more difficult between adjoining freeholders following subsequent transfers of title. This is an over-simplification of a complex problem which has been solved in England and Wales by the practice of granting long leases on flats and apartments rather than selling with a freehold interest. Commonly, 99-year, 125-year or 999-year terms will be encountered, the latter being an over-optimistic figure to apply to the life expectancy of the structure, especially in the case of some post-War speculative flat developments which will be fortunate if they see the end of a 99-year term!

There are two important points affecting flats and apartments which have significance:

(1) The design of the building with particular reference to sound-proofing, and
(2) the means of escape in case of fire.

These are of profound importance to the client if he is to enjoy quiet occupation and physical safety. The survey report should always set out to cover sound-proofing as far as possible and if a noise problem seems likely, the client should be warned.

An elderly couple retiring to a ground floor apartment and hoping to end their days in tranquility will be bitterly disappointed to find that every sound made by a young family in the apartment above can be heard, perhaps late into the night.

Fireproofing between units and the means of escape in the event of fire are vital matters. The standards to apply would be those of the current Building Regulations and also those which would be required before a fire safety certificate would be issued for a new building. If an existing building should fail to meet such standards, the client must be warned and recommendations made. Often, it would be impossible to meet modern requirements in an older building without the full co-operation of all flat owners, and this would be unrealistic. In such circumstances, all the surveyor can do is advise and warn and then leave it to the client to decide whether or not to proceed with the purchase.

Types of flat

In large conurbations, especially in London, there is a ready market for converted flats which began life as conventional houses or other structures and which have been subsequently sub-divided, often to a poor speculative standard. Caution must be exercised in reporting upon such property as there are several potential pitfalls, both physical and legal involved, about which the client will need to be advised.

Generally speaking, in England and Wales there are three main types of flat or apartment development:

1. Those in the form of simple low-rise structures, either purpose-built or converted, consisting mainly of two-storey buildings similar in appearance to two-storey dwellinghouses where separate units of accommodation are provided on each floor and where each unit has its own exterior entrance door. There may be a shared front porch or perhaps one doorway in the front elevation and one at the side. Such accommodation, in estate agents' parlance, is generally referred to as 'a pair of maisonettes' (in the London area at any event). Generally, garden areas are also divided and each unit, together with its garden is held under a separate long lease. Commonly each leaseholder is liable for the interior repair and decoration of his or her own unit. For carrying out repairs to the exterior or structure of the building, a variety of arrangements may be provided, these depending upon the date of the leases since conveyancing fashions have changed over the years, although generally speaking such arrangements fall into one of two broad types. Each lessee may be liable for his or her own part of the building, having no responsibility for any other part. A lateral division of responsibility is provided for the internal structure, this generally being the base of the first floor joists with a proviso that the staircase to the upper unit is included with that property. Alternatively, lessees may be jointly responsible for repairs to the foundations, exterior, structure and roof so that in the event of a repair being required, agreement would have to be reached on the need for and the cost of the repair. Each party would then contribute towards the cost in proportion to their holding.
2. Next is the modern purpose-built block of apartments, generally of post-War origin and with communal grounds, staircases and entrance halls. Occasionally services such as porterage, hot water and central heating are also provided on a communal basis. In such cases there will generally be a liability for each lessee to

maintain the interior of his or her own unit and contribute a service charge for the maintenance and redecoration of the exterior, upkeep of the structure and roof and towards the communal services. Responsibility for arranging the communal services may vest in the freeholder who will then collect a ground rent, service charge and (usually) the insurance premiums due from the lessees, and arrange for the necessary periodic redecoration and upkeep of the exterior and structure of the building, its grounds and the communal services.

Alternatively, the responsibility for maintenance could vest in a separate body which may be a residents' association with a chairman, secretary and treasurer, being subject to rules of procedure and with published annual accounts. Such an association will balance accounts by levying on members a service charge which is sufficient to cover the outgoings. The effectiveness of a residents' association will depend not so much upon the specific rules governing proceedings as on the quality of the residents who may be persuaded (or more usually cajoled) into serving. A well-run residents' association which maintains carefully the communal areas and grounds and ensures a high standard of maintenance and decoration of the buildings is the ideal. Occasionally, indifference on the part of residents or clashes of personality within an association will produce an unhappy effect, some members being anxious to maintain high standards and others clamouring to keep down costs. The standard of maintenance will then suffer considerably.

3. The third type is the older mansion block in larger cities, originally built for letting by private landowners or institutions and having been subsequently subject to 'break-up' operations involving the selling-off of individual units on long leases. Often, original tenants remain in such blocks so that a mixture of tenures is present, some residents having bought long leases and others occupying flats under regulated tenancies protected by the Rent Acts. In such cases either a freeholder will be liable for exterior and structural repairs with a right to recover service charges, or the freeholder would have created a residents' association of some type and be in the process of devolving responsibility for such matters.

Service charges

In preparing a report on a flat or apartment held on a long lease, the surveyor should first read the lease (see Chapter 16), and if the lease provides for a service charge to be levied, he will wish to look with particular interest at the supporting provisions. There are four main requirements for a reasonable service charge provision, these being:

(1) The period of payment should be given, i.e. whether quarterly, half-yearly, yearly or on some other basis.
(2) The items of maintenance, repair, decoration and upkeep, together with all services intended to be covered, should be set out clearly.
(3) The manner in which service charges are apportioned between separate units should be fixed. Probably a number of identical flats in a block will pay the same amounts but flats of different sizes may need to be differentiated by, for example, rateable value, and other modifications may be needed in the interests of fairness (e.g. the ground floor lessees should not pay for lift costs).

(4) The manner in which the maintenance accounts are to be kept and certified should be made clear.

Commonly, where a service charge is provided for in a lease, the freeholder will wish to ensure that all possible liability is devolved upon the lessees and that no residual liability for extraordinary items of expenditure remains with the freeholder. As a consequence, those parts of the provisions dealing with requirement (2) above will often be very widely drawn and embrace all conceivable heads of expenditure. There may even be provision for additional matters to be added in the future.

There is a long history of actions in law and statutory intervention as a result of disputes which may arise between lessors or residents' associations on the one hand and individual lessees on the other over the reasonableness or otherwise of service charges. In *Finchbourne Ltd.* v *Rodrigues* ([1976] 3 All ER 581) the Court of Appeal held that a term should be implied into a service charge clause that costs recoverable by the lessor from the lessee should be 'fair and reasonable' in the circumstances, and also that where the lessor and the managing agents for the property are effectively one and the same, the managing agents will not be sufficiently independent to certify the service charge under a provision for certification.

Section 91a of the Housing Finance Act 1972 was introduced in an attempt to assist lessees and was subsequently modified by Section 124 of theHousing Act 1974 and Schedule 19 to the Housing Act 1980. This provision provides that a service charge shall only be recoverable from the tenant of a flat for the provision of chargeable items to a reasonable standard in line with *Finchbourne Ltd.* v *Rodrigues* (supra), and also that 'to the extent that the liability incurred or the amount defrayed by the landlord in respect of the provision of chargeable items is reasonable'. Provision is also made that in building works costing more than £250, at least two estimates must be obtained by the landlord, one at least of which is from a contractor wholly unconnected with the landlord or the managing agent. In respect of expenditure exceeding £2000, there is a procedure for the landlord to consult with the tenants except in cases of emergency and only then to obtain estimates for the work.

Section 90 of the Housing Finance Act 1972 is designed to enable tenants to obtain information about service charges even if their leases conferred no such right. It may well be that in order to complete his report that surveyor will need to inspect the management accounts. It is unusual to find that a prospective purchaser or his surveyor or even his solicitor have a statutory right to do this but the vendor of a long lease will have such a right, either by virtue of the provisions of the lease itself or by virtue of Section 90. If the surveyor is denied access to this information by the vendor, or if it is suggested that up-to-date accounts are not available, then this must raise serious doubts about the present state of management accounts and the matter may need to be referred back to the client's solicitors for further instructions.

If an inspection of the management accounts has not been possible, the surveyor will be unable to express any view on the reasonableness (or otherwise) of the service charge currently payable, or the likelihood that a substantial future increase in the service charge may be possible because of past under-provision for some items.

When an inspection can be made this should preferably be of accounts over a period of more than one year. Ideally, the last three years' accounts should be scrutinised. In order to forecast future trends, these accounts should be considered in the light of the need to maintain some reserve for periodic decorations, often necessary at three-yearly intervals, this causing a sudden leap in the service charge for one year if an adequate float has not been built up in the two preceeding years. Similarly, special attention should be given to checking the provision for capital replacement of items such as lifts and fixed plant which may have a well-defined life-span. If there are fixed-term contracts for cleaning or gardening and if these are about to expire, a large increase in such costs may arise if an alternative contractor has to be found.

An inspection of a single flat in a block of flats for a prospective purchaser must, of necessity, include not only the flat in question but also the building as a whole since a sudden large increase in the service charge could arise in respect of a defect in the building quite unconnected with the flat under review. In our experience this will often arise in the case of major structural defects to foundations and roofs. It is essential for the client to realise that he may be proportionally liable for this type of defect even if it is not related to the flat he proposes to buy. Similarly, a prospective mortgagee will wish to know about such matters and will expect the building as a whole to have been inspected by the surveyor in addition to the specific flat used as security.

The RICS Practice Note, 'Structural Surveys of Residential Property' states that 'Particularly onerous repair liabilities may exist quite independently of the subject property, such as where a lease or agreement imposes a liability to pay a proportion of the total estate repair costs', and 'To advise fully the surveyor must endeavour to establish the extent and nature of the repairing liabilities under the lease, the client's potential responsibility for executing repairs and also his liability to pay for repairs executed by others'.

If a block of flats suffers from a serious structural defect, the purchase of one of the flats, even where unrelated to the defective area, may be unwise unless all the implications are fully understood. It may be, for example, that a foundation defect exists and that this should be remedied by the National House-Building Council Certificate procedure, or works may be covered partly by the subsidence and landslip provisions in the building's policy. In such a situation, there may be a number of parties involved in a liability dispute including the lessees and their solicitors, the original contractor, the NHBC, the building insurers (possibly under individual policies on the flats) and the local authority who passed the work originally. There may also be a residents' association, and an owner of adjoining trees, the roots of which could have contributed to the defect.

All surveyors will be aware of problems associated with flat roofs, especially **those of poor construction dating from the nineteen-fifties and nineteen-sixties.** Many such roofs were used on blocks of owner-occupied flats and the roof coverings are now nearing the end of their useful life span. Renewal will often be a question not merely of stripping off existing built-up felt or asphalt and re-surfacing. Frequently it will also involve repairs to the decking and supporting construction involving major expenditure on the works and on protection of property and occupiers for the duration of the works.

A major item of expenditure such as renewal of a flat roof covering, presents particular problems to a residents' association or freeholder collecting periodic service charges. Ideally, a contingency fund should have accumulated to pay for this work but this will rarely be found to be so. It may be that the nature of the remedial works required will have to be agreed (itself not an easy matter) and then estimates obtained and contractors instructed on the basis that a sudden large increase in the service charge will have to be made to cover this item.

The chairman of a residents' association in such a situation will feel like the messengers of old who were praised if they brought good news but put to death if the news was bad. The owner of a flat in such a block may actually decide to sell and move just to avoid problems of this type which may be approaching and the purchaser will need to be warned about such circumstances.

Tenure in Scotland compared

In Scotland a different law applies to the sale of flats and apartments. Scottish law is based upon the original feudal tenure modified by statutes, with all rights to land originally derived from the Crown. Buildings and landed property are classified as 'heritable property' and being either 'corporeal' if land or buildings, or 'incorporeal' if merely rights of land or buildings. In these respects the system is similar to that operating in England and Wales. One difference is that a system of charges may be levied, similar to rent charges on freehold land in England and Wales, and termed 'feu duty', normally running in perpetuity with the land, the duty being paid by the owner or 'vassal' to the person or 'superior' who owns the right to receive the duty. Occasionally a superior may have a right to acquire the vassal's interest by virtue of a clause in the title which, in effect, gives him an option to purchase, generally at a price equal to any other offer which the vassal may have received for his interest. Such a right to pre-emption normally arises only on the first sale of the vassal's interest and not subsequently. A similar situation arises in England where an original freeholder may impose restrictive covenants on subsequent purchasers of the freehold interest, although a right to buy back the freehold is less usual.

A wide range of feuing conditions may be imposed on vassals by the superior and may generally be enforced, not only by the superior but also by other vassals who hold parts of the same original holding. This is similar in many respects to the estate covenants which may be imposed by freeholders of estates in England and Wales whereby a series of mutually enforceable covenants may apply to the estate as a whole. This is a particular feature of some residential developments built between the Wars in the higher price ranges where obligations may arise to maintain private roads, open spaces and the like with limitations on the type of alterations or uses permitted.

In English law there must be privity of contract between two parties before action may be taken, but in Scotland a vassal or 'feuer' may take action against another without there being privity of contract between them provided that the aggrieved feuer is personally affected by the breach of feuing conditions and that the superior has not himself permitted the breach specifically.

Against this background of mutually enforceable feuing conditions, which arise irrespective of whether or not there is a direct contractual relationship, it has been possible for property in Scotland to be sold as 'flying freeholds' subject to the feu duty and to the conditions. Scottish law also permits servitudes (the equivalent of English easements) in the form of rights of support, protection and light.

During the course of inspecting a flat or apartment in Scotland the surveyor will need to obtain details of the feuing conditions set out in the title and consider whether or not these are adequate and suitable. In so far as potential liabilities may arise for the maintenance and repair of either the structure as a whole or other parts of the building separate from the specific flat concerned, these may need to be investigated and the position explained to the client.

Conversions

Apart from general distinctions of tenure and various types of flats and apartments which may be encountered, there are two broad divisions of type which need to be identified. These are (a) those flats or apartments which have been purpose-built, and (b) those which are conversions. The surveyor should make a particular point of confirming in all cases which of these two situations applies since this is relevant to the advice which may be given and may be significant to a prospective mortgagee or client proposing to raise a mortgage. Institutional lending policy on converted flats is often very different from that applied to purpose-built flats.

In most cases the converted flat will be fairly obvious. An original Victorian or Edwardian villa in a city suburb may have two entrance doors in the porchway and a plasterboard and stud partition wall dividing the original hall. The nature of the original house and the extent of the conversion works will be self-evident in such circumstances.

Faced with such a conversion, the surveyor should see if the staircase to the upper flat has the necessary degree of fire-resistance and that the partitioning and undersides of the staircase are suitably fire-resistant, at least up to the standards of the current Building Regulations or London Building Acts. Similarly, if the upper floors are simply suspended timber floors, these will have required additional fireproofing to meet regulations. The reason is to ensure that a fire in the lower flat will not spread rapidly to the flat above and that in the event of fire, the occupants of the flat above may have a protected means of escape down the stairs.

Many converted flats have never been subject to any scrutiny by a building control officer or district surveyor. The reason for this is that they have been 'converted by stealth'. Probably a house was first used as accommodation for two families. Later, various modifications were perhaps undertaken, culminating in sub-division of the accommodation and splitting of the services. If the change of use to two units of accommodation is of long-standing and if internal alterations are purely internal, no reference to the local planning department may have been made. Finally, the tenants leave and the owner, much impressed by the current level of property values for converted flats, will decide to sell the house in two parts by creating two separate leases. At this stage, one of the purchasers commissions a survey and finds that there are wide-scale breaches of the Building Regulations or other regulations. If the fireproofing between units is affected, the building may actually be dangerous. The surveyor will often find partitions of plywood or hardboard masquerading as plasterboard. Often various defects such as rising damp, decay to timbers, defective ceilings and roof problems will have been dealt with in a highly unsatisfactory manner and a variety of defects may have been plastered and decorated over.

Occasionally it may be difficult to confirm initially whether or not a particular property has been converted, and some enquiries regarding its history or exposure of the structure could reveal the truth.

Chapter 14

Reports on new buildings

Where there is a contract to build and sell a house or to sell a partly-built house when it is completed, the law makes a three-fold implication: that the builder will do his work in a good and workmanlike manner; that he will supply good and proper materials; and that it will be reasonably fit for human habitation.
Hancock v *B. W. Brazier (Anerley) Ltd. (1966)*

National House-Building Scheme

The National House-Building Council's *Registered House-Builders' Handbook* is essential reading for any surveyor dealing with new houses and flats. This is a first class and easily understood guide to NHBC standards which covers all the main topics likely to arise and is written in a simple straightforward manner.

Should any new residential property fail to meet the standards laid down by the NHBC then this should be commented upon in the report. Even where a NHBC inspector has passed the building minor technical breaches of the *Handbook* requirements will often be found as, due to insufficient staffing, the NHBC cannot inspect every part of every new house during every phase of construction.

The procedures for taking instructions and carrying out the inspection, also the preparation of the report, will generally be the same as those for older buildings, but certain special considerations apply to new property.

First, the client will have a contract with the builder or developer and will need to know whether or not the requirements of that contract have been fulfilled. Even if the contract is vague about the standard of construction and finish to be provided, the law makes certain stipulations. Where the property is residential, the requirements of the NHBC should be met unless no NHBC certificate is to be issued (which nowadays is an unusual situation and one which would call for special advice to be given). In the case of a 'second-hand' house, the contract between vendor and purchaser will be one which will give the purchaser little remedy against the vendor covering defects in the property being conveyed, although a property less than ten years old may enjoy the benefits of the residue of a ten-year NHBC "Buildmark" certificate in respect of certain types of major structural defect.

Secondly, because the structure is new no advice can be given to the client about how the building may have performed for a period of time. A second-hand house which is a few years old will normally start to exhibit signs of structural distress if there is some fundamental fault in its structure or in the underlying subsoil. Generally, it is recognised that such major faults almost always become apparent during the first five years of a building's life.

Because the structure cannot be subject to an extended period of observation and therefore will not have proved itself stable by the passage of time, the surveyor will normally make enquiries over the original design and will have to inspect working drawings, plans, engineer's calculations and so forth, in order to assess the quality of that design.

Normally, the builder or developer would make available any plans and calculations for the surveyor's inspection. If permission to scrutinise such details is refused, it is felt that the surveyor might well then decide to refer the matter back to his client on the basis that in such circumstances he cannot prepare an adequate report. A telephone call to the district surveyor or building inspector is often advisable. It is surprising to find how often a notionally completed building has still not been finally passed under the Building Regulations or other legislation due to some breach of these Regulations or other problem of design or construction. Clients should be advised not to complete a purchase until all necessary final inspections have been undertaken and the building is finally approved.

Planning and other permissions

Another important point to consider in any new building is that almost all new construction in the United Kingdom is subject to the requirements of various permissions and consents and buildings erected without all the necessary permissions and consents may be illegal or unlawful. Clearly, before a client buys or leases a new building and before a mortgagee lends money on the security of a new building, the fact that all necessary consents have been obtained must be confirmed.

In the first instance, it is the client's solicitor who is responsible for finding out whether or not the new building exists legally and lawfully and that there are no breaches of the law. As has been mentioned before, however, it is the surveyor who actually looks at the property – the solicitor will rarely see the property he conveys for a client so the advice which the surveyor can give his client or his client's solicitor will prove to be important.

Basically, there are four types of consent which may be required for new buildings:

1. Permissions under the Town and Country Planning Acts, applying in most cases but perhaps not in the case of certain types of small extensions to existing buildings or to exempt buildings such as those used for agricultural purposes.
2. Permission under building control codes, quite a separate system from planning and governed by the Building Regulations 1985, which were amended as from 1 April, 1990 in respect of approved documents F, G, H, J and L and which provide the basis of building control in England and Wales. Similar regulations apply in Scotland and Northern Ireland. Building construction legislation is subject to periodic review, and it is important to keep up to date.

3. The consent of a landlord or freeholder of leasehold property; of an original freeholder in England and Wales who holds the benefit of restrictive covenants; or, in Scotland, the superior to a title who holds the benefit of feuing conditions. Examples may be found where titles include old restrictive covenants against redevelopment at higher densities or different types and such conditions may be removed only by applications to the Courts or Lands Tribunal. Specific insurance cover may sometimes be available against the possibility that an old restrictive covenant could be enforced and building works may then proceed with the benefit of such insurance cover. If title insurance is relied upon, the surveyor should obtain details and confirm whether or not the amount of cover is adequate since it will often be found insufficient to compensate fully a purchaser in the event of a restrictive covenant being enforced.
4. The consent specific to the particular type of building having regard to particular local or national legislation. Special rules apply to filling stations and garages which require petroleum licences and if a petroleum licence is refused or revoked, this could have a disastrous effect on the value of the premises. Similarly, fire safety certificates are required for hotels, public buildings and the like. It is important to confirm that a new building meets all such requirements, including projected changes of legislation which may arise in the future. Offices and shops are governed by the Offices, Shops and Railway Premises Act and factories by the Factories Act. In the case of buildings where people will be employed, the Health and Safety at Work legislation may apply to certain aspects of new building design.

If a new building is not residential then the surveyor will consider whether or not the design complies with all existing and projected local and national legislation to which it will be subject, having regard to the purpose for which the building was designed and the purpose for which the client proposes to use it, taking into account any potential change of use.

The fact that planning permission and approvals necessary under building control legislation have been granted does not ensure good design. The fact that a new building may have been inspected periodically by a building inspector or district surveyor does not ensure good standards of construction since such inspections, by their very nature, can only be 'spot checks'. Neither will the fact that a new house may have been inspected by the NHBC indicate good standards of construction, or even that all NHBC requirements have been met, since these inspections also are only spot checks.

Therefore, when reporting on a new building, the surveyor should not accept anything at face value, neither should he rely upon the past supervision or inspection by others as an indication that a particular item is satisfactory. He can only report properly on those matters which he can see or discover for himself.

Foundation design and subsoil

The first subject for his personal investigation is the nature of the supporting subsoil and the foundation design. In this connection, the architect's or builder's design drawings as approved by the local authority will at least show the intentions. Consideration should then be given to the question of whether or not a design is suitable for the indicated subsoil conditions and enquiries made to discover what subsoil samples were taken and whether or not any trial bores were made on the site and to what depths. It may be that results of a laboratory examination of soil samples can be made available. If not, and if the position seems doubtful, it may be that the surveyor himself will need to recommend that his client commissions an analysis of the subsoil. In recent years, however, most local authority building control departments have required a note to be made on the submitted drawings to the effect that the foundations will be designed to meet site conditions rather than requiring an explicit detail on the drawings.

The reason for this emphasis on the foundations and subsoil under a new building is that if a building has not proved itself by performance over any extended period of time, the danger, albeit remote in most cases, is that the foundations may be inadequate or the subsoil unstable. From a client's point of view, the whole point of instructing a surveyor in the first instance is to receive advice regarding just such matters and a report which merely states that the building has not been subjected to any extended period of observation and that no advice can be given on the suitability of the foundations will seem distinctly unhelpful to a client. He will then wonder whether the exercise of instructing a surveyor was really worthwhile.

Certain obvious warning signs which should place a surveyor on his guard over a foundation design would include the following:

(1) Any known history of mining in the locality.
(2) Any known history of sand, gravel or chalk extraction or filled gravel pits etc. in the locality.
(3) Sites with steep slopes, retained by embankments or retaining walls, or where adjoining structures or highways show signs of landslips.
(4) Any indications of tipping or landfilling in the locality.
(5) The presence of shrinkable soils, especially near vegetation.
(6) Any possibility of coastal erosion.
(7) Signs of structural distress to older buildings in the locality.
(8) Trees closer to the structure than their mature height and which could cause subsoil shrinkage. Evidence of the recent removal of trees from a site where ground heave is likely.
(9) Adjacent flat, marshy or poorly-drained land.

Another useful indicator is to make a check on the identity and qualifications of those responsible for the foundation design. Generally, it is to be hoped that consulting engineers were employed to report on soil conditions, prepare structural calculations and designs and that qualified architects or surveyors were responsible for working drawings and layouts. Often, however, drawings are prepared by 'in-house' personnel employed by the builder or developer.

Occasionally drawings will be found carrying no indication as to the name or qualifications of the draughtsman. No hard and fast rules can be given for the acceptability or otherwise of designs prepared by unqualified or unknown personnel since they may well be experienced and competent. However, the lack of any outside professional involvement in structural design can be an indication that it was felt necessary to avoid payment of professional fees and cost cutting may then also be evident in the design or the construction work itself.

It is sometimes helpful to talk to the men working on a building site if this is possible and it is surprising how much can be learned from such casual conversations. They will give an indication of the loyalty of the workforce and the efficiency of the employer in providing adequate supervison. They will often readily discuss problems which they have encountered on site, difficulties with the local authority or faults and poor deliveries of materials. An overall impression of the general standard of care being taken with the building work can be gauged. On no account, however, should members of a workforce be 'quizzed' on the progress, administration and finances of a contract.

Inspecting the building

Having completed all the investigations which are possible and are required by the client in respect of subsoil and foundation design, the surveyor will then proceed with his inspection of the building. This inspection will follow the same lines as the inspection of any other building using the surveyor's own normal routine. If the building work is incomplete, it is suggested that a list of all unfinished items be prepared and included in an appendix to the report with a note that the surveyor, if required, can confirm that such matters have been completed at a later time.

At each stage of the inspection, all defective work and deviations from good building practice should be noted. In the case of a residential property a check should be made for breaches of building control legislation applicable to the locality and any lack of conformity with the standards laid down in the NHBC's *Registered House-Builders' Handbook*. In the case of non-residential property, the check should be for breaches of building control legislation and any other lapse from the specific requirements of that particular building type.

Surface finishes must be checked with particular care, especially in the case of a residential property. Clients buying new houses are doing so for many reasons, but one requirement will be that the decorations will all be new. The clients will then expect that redecoration will not be necessary for several years, this clearly being a saving compared with the typical situation in a second-hand house where most buyers expect to have to redecorate in accordance with their own tastes soon, if not immediately, after moving in. Experience shows that purchasers of new houses tend to complain most over problems of final finishes to walls, ceilings and woodwork and by lack of care when installing fixtures and fittings. It is necessary therefore to apply a stringent standard to a new house where these matters are concerned, as compared with a second-hand house where most clients will accept a lower standard knowing it is their intention to improve to their own tastes. Having said this, however, clients should be told that a builder's initial finish is not as durable as later-applied finishes may be, especially in the case of new painted softwood.

During the course of a new building inspection, an opportunity arises for the surveyor to consider the overall quality of design from the client's point of view and to comment unfavourably where necessary. In commercial and industrial structures the durability of finishes is important in view of the considerable wear and tear which may arise in connection with the client's trade or business. Floor finishes in commercial and industrial structures are very important from the point of view of hardness, freedom from dust, ease of cleaning and non-slip surface. Often, hazardous floor finishes or potential dangers in a building will be encountered which may possible give rise to claims from injured employees or visitors. Clients will expect to have been warned of such problems by their surveyor in advance.

Security in a new commercial or industrial building may be poor and a client may have to undertake protective improvements before being able to obtain insurance cover, especially if warehousing is involved or valuable goods stored. Often new factory and warehouse space is built and let, or sold, as a shell on the basis that the purchaser or tenant will undertake his own fitting-out. The provision of adequate security, fire safety precautions and fitting-out to comply with local and national legislation will have to be allowed for by the client and the advice of the surveyor in budgeting for these costs may be relevant.

The quality of fittings provided in new speculative buildings may be poor. In dwelling construction there is a wide range of qualities available in items such as sanitary ware, kitchen units, fenestration and door and window furniture. The temptation for the builder or developer is to use fittings which are superficially attractive but less durable, especially in the case of sanitary ware. A client may see a new and attractive bathroom and be impressed by the appearance of the sanitary ware and fittings, without being aware of the distinction between acrylic, steel and cast-iron baths, or the variations in enamel thickness, or the tendency for acrylic to flex in use and suffer surface deterioration from abrasion. At the present time for example, it is possible to buy baths at prices which, for the most expensive, are ten times the cost of the cheapest with very little difference in superficial appearance.

The internal planning of new accommodation is important. In residential construction it is necessary to consider whether or not the rooms will be of suitable shape and size to accommodate normal furniture satisfactorily and if furniture can be conveniently moved in and out. For example, to be usable bedrooms must be of a size and shape to accommodate a bed while leaving adequate access for bed-making and space for the door to be opened. (This may seem obvious, but there are many instances of small bedrooms being virtually unusable for this reason).

Suitable access for reading electricity and gas meters and provision for the storage of rubbish bins are necessary. External meter boxes are clearly preferable and should be located in front of any lockable side gates so that access from the highway is always free. Similarly, rubbish bins should be kept out of sight but convenient for use and for emptying. Correct planning of water-closets and bathrooms is necessary to ensure adequate room for the convenient use of appliances and door opening. Convenient kitchen layouts must avoid excessive movement being necessary from one area to another in kitchen operations; a layout of worktop – cooker – worktop – sink – worktop – fridge/freezer is preferable, unbroken by doorways or traffic-ways.

Special attention should be given to staircase and balustrade design, this often being poor. Widths should be sufficient to allow for furniture removal; handrails should be adequate and extend for the full length of the flight; good lighting is necessary, and balustrades should be provided with gaps such as to prevent the smallest child from falling through or becoming jammed. The Building Regulations demand that the space between balusters should be such as to prevent the passage of a 100-mm diameter sphere.

Recent evidence tends to suggest that the overall quality of new building design and construction has improved during the 1980s and that lessons have been learnt from the failures of the previous two decades with a trend back to time-tested traditional methods.

Chapter 15

Reports on older buildings

Built in the old Colonial day,	With weather stains upon the wall
When men lived in a grander way,	And stairways, worn and crazy doors,
With ample hospitality;	And creaking and uneven floors,
A kind of old Hobgoblin Hall,	And chimneys huge, and tiled and tall.
Now somewhat fallen to decay,	*Longfellow*

Since it is most unlikely that any surveyor would need to report on a building dating from before the Early English period, I propose to deal with post Early English construction only.

When stone was the principal building material used in church and cathedral construction, for example, materials in areas devoid of roads, such as the fens, had to be conveyed on horseback and in some districts, transported by sea or river. From the earliest times, therefore, bearing in mind the primitive or absent transport, local materials have been used as close to the site of origin as possible. The limestones of Portland, granites of Aberdeen and Cornwall, Bath stone – all were first used locally, but as transport became more efficient, local distinctions disappeared and even in the Middle Ages, stone began to be brought from a distance such as the Normandy Caen stone used in Canterbury Cathedral.

However, for less grand buildings, locally-available materials were used principally in domestic construction right up to the 20th century.

The modern type of brick, for instance, came into use around AD 1300, having lapsed since Roman times, and the earliest brick building which survived into the 20th century is Little Wenham Hall in Suffolk (AD 1260). The predominance of brick for domestic construction dates from the reign of William and Mary and, broadly speaking, it is still the most economic material for external facings to properties of all sizes.

Half-timbering as in Lancashire, Cheshire and the West Midlands dates largely from the 14th to 16th centuries, terra-cotta in Essex was used in the early 16th century, and in chalky districts a great many medieval buildings were constructed in flint, still common in Norfolk, Suffolk and Hertfordshire.

The variety of structural materials encountered in the older buildings is therefore perhaps even greater than that of 20th century construction.

In both ecclesiastical and secular building, the earliest periods are:

1189–1272 Early English
1307–1377 Decorated
1377–1558 Perpendicular

All these three periods represent the Gothic style in Great Britain and, in the main, survive as ecclesiastical buildings of primary concern to those who carry out quinquennial surveys of churches and are largely of academic interest only to the general practice surveyor.

Of greater interest to surveyors is the Renaissance period which commenced at the start of the 16th Century after the Wars of the Roses and during the suppression of the monasteries (1536–1540). These events resulted in a revival of architecture, the conversion of some ecclesiastical buildings into mansions, the endowing of grammar schools and, during the prosperous Elizabethan period, a great increase in the building of country houses such as Longleat House, Wiltshire, and Haddon Hall, Derbyshire.

The Jacobean period saw such mansions as Holland House, Kensington and Bolsover Castle, Derbyshire, completed during the reign of James the First. These and the preceding 'Elizabethan' mansions were principally of brick or stone but timber construction was still employed in Cheshire, Lancashire and Shropshire. Many of the Oxford and Cambridge colleges have buildings belonging to this period such as Emmanuel College, Cambridge, and Jesus College, Oxford.

The Anglo-Italian period was introduced during the reign of Charles the First who employed Inigo Jones (1573–1652) to design complicated structures of classic grandeur. Sir Christopher Wren (1632–1723) also came within this period and we are familiar with his works but, although a genius in construction, he did not have the decorative flair displayed by Inigo Jones.

From here there was a logical progression to the 18th Century styles which split naturally into two periods:

1702–1714 Queen Anne
1714–1820 Georgian (the reigns of the first three Georges).

During the former period, Castle Howard and Blenheim Palace were designed by Sir John Vanbrugh in the lavish Anglo-Classical style, although minor domestic architecture during Queen Anne's reign was generally plain and functional with much brick used in construction – a wonderful period for domestic design.

In the latter period, Robert Adam (1728–1792) was probably the most famous exponent of the Georgian revival, as witness Adelphi Terrace, London, and Keddleston Hall, Derbyshire. Columns, vaulting and sprawling plans again became fashionable with strong emphasis on façades. This prompted the Classical Revival with ambitious pseudo-Greek models as in the Bank of England (1788) and St George's Hall, Liverpool, by H. L. Elmes (1815–1847).

An upsurge of spiritual activity in the Church then brought us to the Gothic Revival in the 19th Century with Augustus W. N. Pugin (1812–1852) as the apostle of ecclesiastical work in this style. The most impressive buildings of this time are, however, the Houses of Parliament designed by Sir Charles Barry with Pugin's help, and the New Law Courts designed by G. E. Street (1824–1881) who was a pupil of the famed Sir Gilbert Scott. Further notable buildings of this period include Keble College, Oxford, Truro Cathedral and the Manchester Town Hall.

About this time commenced the 'Free Classic' movement which attempted to use the best of classical design and construction without slavish reproduction, for example, Norman Shaw's New Zealand Chambers in Leadenhall Street, London.

Although Sir Gilbert Scott, who designed the Home and Foreign Offices between 1860 and 1870, is probably the best-known architect of those times, Norman Shaw, W. E. Nesfield (Regent's Park Lodges) and Philip Webb (Lord Carlisle's house in Kensington) had the greatest influence over domestic-scaled design.

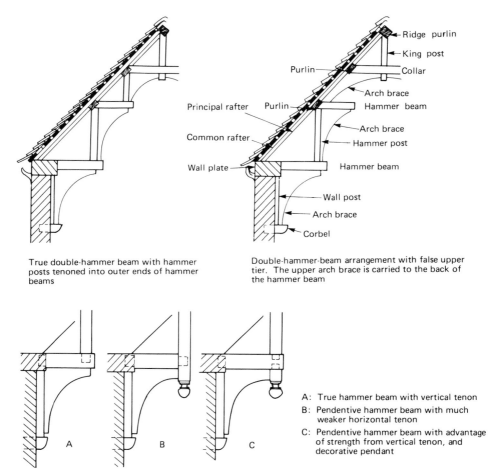

True double-hammer beam with hammer posts tenoned into outer ends of hammer beams

Double-hammer-beam arrangement with false upper tier. The upper arch brace is carried to the back of the hammer beam

A: True hammer beam with vertical tenon
B: Pendentive hammer beam with much weaker horizontal tenon
C: Pendentive hammer beam with advantage of strength from vertical tenon, and decorative pendant

Figure 15.1 Typical hammer-beam construction used mainly in churches and cathedrals. Detailed knowledge of this type of construction is essential before an accurate assessment of strength may be made

This then brings us up to the 20th Century and the substantial and well-founded works of architects such as **Sir Edwin Lutyens RA whose Castle Drogo in Devon,** completed in 1930, was the last great country house to be built in Great Britain. Here, however, experiments in materials – particularly the use of asphalt for flat roofs, and damp-proof membranes in solid granite walls – has led to new disaster and involved the present owner, the National Trust, in much expensive remedial work. His more traditional domestic buildings, however, such as 7 St James's Square, London, and Heathcote, Ilkley, Yorkshire, are well-found and, in addition, a delight to the eye. Other early 20th Century architects of note in the domestic field were Clough Williams-Ellis whose brainchild was the eccentric and fanciful village of Portmeirion on the North Wales coast, and Percy Scott Worthington with his Cheshire 'stockbroker belt' magnificence.

Traditional methods of design and construction

The buildings themselves are even more varied and sometimes more temperamental than the occupiers; some inspire love, others hatred, but they are never dull. A few brief personal notes and opinions on typical differing types are now given to complement the foregoing short history. This is not intended to be a complete treatise on defects to be found in old buildings but just a guide to the faults one may anticipate:

Elizabethan and Jacobean

Depending upon the part of the country in which these buildings are to be found, one can be very badly caught out on occasion by historic mansions and country cottages.

Generally, walls are not plumb and vary in construction from timber lath and plaster via porous brickwork and eroded freestone to distintegrating cob. Roof timbers are often propped with tree branches from which the bark has not yet been stripped. This is indeed fertile ground for furniture beetle infestation and, depending upon geographical situation, properties of this age and type are most prone to this defect. As a great deal of oak and hardwood was also used in their construction, they are unfortunately also liable to death watch beetle infestation.

It goes without saying that floors which are laid in either stone flags or herring-bone brickwork on earth are very damp indeed and the walls are usually prey to rising damp and, in a great many cases, actual damp penetration. In the case of cob-walled cottages, however, the cob *must* be damp otherwise if it is too dry, the walls will disintegrate into piles of dust, having then no moisture content to make the earth filling cohesive.

Queen Anne, Georgian and Regency

These are generally attractive in appearance but often of flimsy construction. Solid walls are frequently not thick enough to keep out the dampness caused by driving rain, one example being the 150-mm-thick soft ashlar stonework used in the external walling of classical terraced housing, particularly in the Bath area. In these cases, the stonework is often battened off internally and covered with lath and plaster. Sometimes, even the lath and plaster is omitted and hessian used instead, or perhaps again battened and covered with hessian and paper to conceal damp plaster.

In the above periods, one has to watch out for lead flats, parapet gutters and valley gutters which, if not properly maintained, are often found to be laid on rotted boarding. Generally roofs are slated, low-pitched and not underfelted, but possibly have been torched in the past, the torching by now having entirely dropped off leaving the roof covering liable to penetration by driving rain and snow.

Again, no damp courses are present and therefore rising damp is likely to be encountered and, unlike earlier properties, there is a very good chance that dry-rot

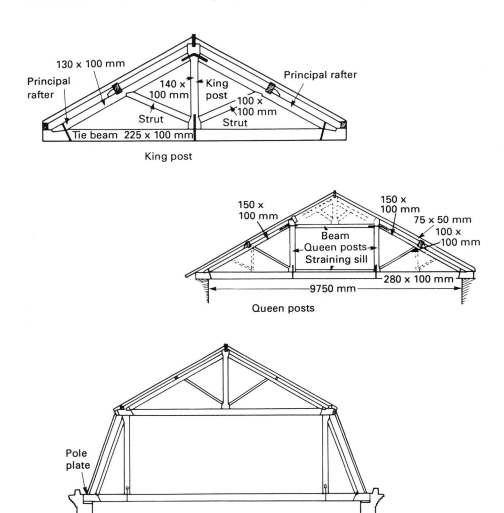

Figure 15.2 Early timber trusses, variations of which will be commonly encountered in older buildings

may be found in the timbers, ventilation usually being inadequate. However, one bright spot is that roof timbers are usually sound with kingpost or queenpost roof trusses, although again liable to furniture beetle infestation depending upon geographical location.

More often than not, ground floor joists are laid directly on to the earth and almost always sub-floors remain insufficiently ventilated.

Later Victorian and Edwardian

These are usually well-found, ugly, solid buildings but often, particularly in the case of those having stone facings, having too much fussy detail which collects pollution of all kinds, creates maintenance problems and very often weathers badly. Later buildings in this category may have slate damp-proof courses in the walls but the normal troubles found in construction carried out during this period are dry-rot, particularly in cellars, and furniture beetle infestation in roof and floor timbers. Once again, walls were built solid during this period but generally more substantial than in earlier properties.

This, in some cases, can create damp penetration problems due to lack of a cavity, particularly if the exterior rendering, applied originally as a waterproofer, has passed its useful life and become off-key and porous, as is often the case in the West Country.

The nineteen-twenties

These are generally good sound buildings, some with cavity walls, damp-proof courses (often slate) and well-constructed pitched roofs of good quality slate or clay tiles. Generally very little trouble is experienced from buildings erected during this period provided that they have been well-maintained during their lifetime. However, flat roofs were often covered in asphalt which by now is sure to have passed its useful life span in spite of maintenance. Wall ties which were not galvanised may now have rusted through, especially in exposed situations, leaving two separate single-brick leaves in a rather unstable wall. Tile coverings may be approaching the end of their useful life. Watch for rusting metal windows.

The nineteen-thirties

On the domestic scale, houses built during this period were very often flimsy, 'jerry-built' and, being the age of the speculative house, they were very often of poor construction, and a quick turnover for large profits was the order of the day. During this period, bituminous felt damp courses were used and now, 50 years later, these are often found to have perished, become brittle and broken down. Again, investigation will sometimes reveal that the early metal ties in the cavity walls can have rusted through after such a period of time since some of the original supposedly galvanised metal cavity ties had not been properly treated. Walls are then left in the state of two separate unconnected leaves of brickwork, often unstable.

Unseasoned roof timber was occasionally used which resulted in warping and distortion plus the possibility of furniture beetle infestation since generally roof timbers were not treated and there was a high proportion of sapwood present.

The nineteen-fifties

Buildings erected during this period were generally simple, solid, built to a budget and on the whole unattractive, but reasonably soundly constructed. There were, however, felt damp-proof courses which, as stated above, have a limited life, there was no cavity insulation to the walls, most ground floors were of concrete, because of restrictions on the use of timber, and many of these had inadequate damp-proof membranes. Floor coverings often consisted of Magnesite composition, early thermoplastic tiles or woodblocks. Magnesite composition has a very limited life and will disintegrate after a few years of heavy use; the early tiles were brittle and broke easily, particularly across the corners; woodblocks, again often of largely unseasoned hardwoods, are liable to attack by various forms of rot. Flat roofs were the fashion and if constructed of timber, very often by this time will have had their problems. Many flat roofs were then covered with asphalt which was unvented, interstitial condensation problems were not considered important at that time due to lower internal heating requirements and therefore by now problems will have occurred in timber roofs covered with boarding and asphalt which are not ventilated and which contain no vapour barriers. Wet-rot to exterior woodwork, especially sills and exposed fenestration, is frequently found.

The nineteen-sixties

During this decade, 'experiment', often wasteful of resources, was the order of the day. New and untried designs, methods and materials were being used. Among designers there seemed to be a strange lack of knowledge about the principles of building construction and many of the new methods which were not tested by time have since failed, sometimes quite rapidly and with disastrous results. This applied particularly to the various novel forms of roof construction which were used.

Once again, domestic construction was becoming 'jerry-built' and flimsy, as in the nineteen-thirties and the use of felt and chippings finishes to flat roofs was probably at its zenith. The motto during this period appeared to be 'live now – pay later' as those who still have to live with high-rise and curtain walling realise only too well!

The nineteen-seventies, eighties and nineties

Both design and construction appear to have improved very gradually over this period. More effective damp-proofing arrangements have been employed, the use of vapour barriers and the problems encountered by condensation were better understood, energy conservation was more fully considered and thermal insulation requirements are higher.

The message concerning the false economy of using flat roofs appears to have been driven home and more pitched roofs are now being employed on all types of buildings. However, now, in the nineteen-nineties, energy conservation methods and the super-insulation of buildings have often had an adverse effect on structure and finishes caused by interstitial and surface condensation problems, lack of ventilation, and a reduction in natural lighting – the result of over-emphasis.

It is worth mentioning that an eminent architect had notices printed which stated baldly, 'All flat roofs leak'. He made each one of his assistants pin up a copy of this notice over his drawing board, a primitive but effective method of performance feedback.

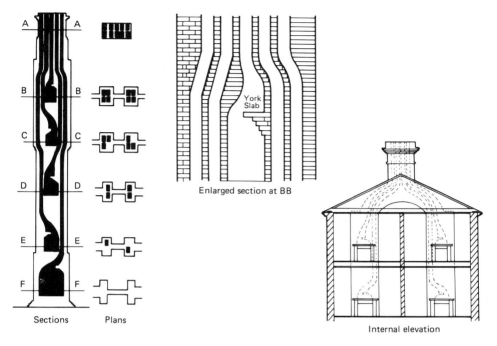

A   A

B   B

C   C

D   D

E   E

F   F

Sections   Plans

York
Slab

Enlarged section at BB

Internal elevation

Figure 15.3 Nineteenth-century flue construction. During inspection it is wise to count the chimney pots and trace the flues within the building

In this country the object of a roof should not be to hold water but to obtain the swiftest run-off possible, and to this end perhaps the steeper-pitched tiled or slated roofs are the best answer.

Local characteristics

When carrying out structural surveys, one can often learn a great deal about how design and construction should fit in with the local environment from both the geographical and climatic point of view. For example, in the Big Apple Restaurant which occupied the 110th floor of the World Trade Centre in Manhatten, attention was drawn to the fact that the management were very worried about cracks which had appeared in the plastered internal finishes to walls and ceilings. Structural failure was suspected by the occupiers but the real inference should be obvious – do not use plastered finishes near the top of a 400-m high building. The amount of 'sway' on this building is bound to make any plastered or rendered finish break down sooner or later.

Again, a textile factory was visited in an exposed situation some 2000 m up in the Colombian Andes where complaints had been made about water seepage through the concrete flat roof over the single-storey office block. This had resulted in much efflorescence and also in small stalactites hanging from the ceilings in many of the rooms. On further investigation it was found that the 225-mm thick reinforced concrete roof had no means of built-in waterproofing whatsoever, neither had it any waterproof covering. The rainfall, heat and humidity in the Andes being what they are, these penetration problems came as no surprise. In addition, the lack of thermal insulation gave the office occupants a hard time under the equatorial sunshine and the answer was finally achieved by superimposing a low-pitched, deep-eaved, corrugated asbestos sheeted roof over the concrete flat, so shedding the water rapidly away from the building and providing a satisfactory degree of insulation from the fierce heat of the sun.

As a contrast, on an inspection of a four-storey dwellinghouse in Turkey on the shores of the Bosphorus, it was found that this property, which was some 120 years old, was constructed entirely of 225-mm brickwork clad on the outside with cedar weather-boarding. The roof was of simple pitched-and-gabled construction covered with red clay pantiles and having very deep overhanging eaves which protected the wall surfaces below from the wet winter weather. Joinery to doors and windows was also in cedar but of a harder and more durable variety than is normally used in this country. None of the woodwork was painted. In spite of this property's repulsive appearance and seemingly flimsy construction, it had stood the rigours of the last 120 years very well indeed and showed no signs of an insect infestation, rot, structural or settlement faults in spite of the fact that one side of the building was literally in the water and the basement contained a boat house from which, under cover, one could step comfortably directly into the kitchen.

The lesson from these examples is that should one see a building constructed of materials which are not indigenous to that particular location and of a design which is not traditional, then one must begin to look very seriously for potential defects. Buildings, of traditional construction using local materials are very often the product of up to 300 or 400 years of experience in performance of such designs and materials for a particular locality. In such cases, practical feedback on performance has probably been handed down over the generations.

To take just a few examples from this country:

(1) *Cornwall*: Walls which appear to be of solid granite in large blocks often consist of two thin slabs of granite laid on edge with a dry earth filling rammed down between. Although the granite slabs may not be more than 150 mm thick, the total wall thickness is then perhaps upwards of 600 mm, which is excellent both from a thermal insulation point of view and for a measure of protection from damp penetration but a problem if one wishes to make an opening in the wall, as one could easily be left with two granite skins and no filling!

The roofs often comprise random 'peggy' slates, the cast-offs from Delabole or other local quarries, fixed with oak pegs and over the years covered with layers of cement slurry. This is extremely effective in preventing driving rain from entering the roofing, but the slurry eventually causes the roof to become a homogeneous slab liable to crack in inconvenient places and suddenly return to earth, in whole or in part.

In central Cornwall, the local stone which is often used unrendered, is of a soft slatey variety which is often laid on the wrong bed. In more than one property it has been possible to scoop stone out of the surface of such walls with a forefinger, and very often the only solution to prevent the wall from disappearing altogether is to cover it with expanded metal, render and roughcast – a favourite treatment for decaying cob walls as well.

It is known that during the nineteen-twenties and -thirties several hundred houses built in Cornwall included Mundic blocks in their construction. These blocks were manufactured from the waste from the East Wheal Rose tin mine at Newlyn, and they contain arsenic, lead and antimony. In time, such blocks suffer complete loss of strength due to reaction of the concrete and sulphides of these metals to form gypsum. Houses built in Mundic blocks can suffer sudden deterioration and even collapse.

(2) *Devon*: West Devon walls, particularly in the Plymouth area, are usually of local stone (limestone in the case of Plymouth) which has been rendered and painted externally. These walls rely on the integrity of the rendering and external paintwork for their weather-proofing qualities. Rising damp is generally prevalent and if the building happens to catch fire, the limestone is calcined and the load-bearing walls may have to be demolished. Roofs are generally of grey Delabole slating or similar, but laid to even courses.

In East Devon, the traditional construction is cob and thatch. These cob walls vary in width, sometimes being almost a metre thick in places. Again, such walls rely on the integrity of the external rendering which, if allowed to crack or disintegrate, allows the weather to penetrate the cob which very soon turns to mud. Usually built on stone pinnings about one metre high.

Conversely, if one attempts to insert a damp-proof course of any kind into a cob wall, there is a tendency for the cob to dry out which means that in the long term one could be left with a pile of dust rather than a wall! Excellent from a thermal insulation point of view, however, and curiously enough not much trouble from rising damp, rot or other like ailments. The thatched roofs seldom have gutters but have a generous eaves overhang, the thatch slowing down the run-off of rainwater and the eaves overhang throwing it well clear of the cob walls. One must not be alarmed by the lumpy and out-of-plumb appearance of most cob walls, but beware of external cracks.

(3) *Somerset*: Somerset is the county of red clay tiled roofs – usually pantiles, double Roman tiles or the Bridgwater interlocking variety. Generally the clay tiles are laid to a reasonable pitch of around 40° and shed the water very quickly but one must beware of pantiled roofs. Very often these are ill-fitting and not underfelted, and one can find some alarming timber structures supporting these Somerset roofs, very often heavily infested with *anobium punctatum*. Clay tiles are now sometimes being replaced by concrete tiles of various forms.

Wall materials vary from local brick to local stone and there are some particularly attractive stones used such as Bath, Doulting and Ham, all with rich colours but often rather friable and, in some cases, with poor weathering qualities. These stones are easily worked and are very often used for the ornamental dressings in some of the buildings, the more local limestone random rubble being used for the general walling with freestone dressings to windows and door openings.

There is a particularly hard stone in South Gloucestershire/North Somerset known as 'pennant' which is often used for paving slabs but can also be used for walling, but beware of internal wet spots since some pennant stone walls may be porous in places in spite of the hardness of the material.

Properties in the Bath area have previously been mentioned but be re-minded of substantial-looking terraces and crescents which on further investigation are found to have deceptively thin walls and are particularly liable to damp penetration, dry-rot, wet-rot and similar ailments which occur when weatherproofing is inadequate and ventilation is insufficient.

One could ponder on traditional methods of construction in various parts of the country for a considerable time. For example, the traditional Lancashire construction of solid millstone walls with outward-sloping joints and millstone flagged pitched roofs; traditional Cheshire construction of half-timbering with timber lath and plaster infilling painted outside and again flagged roofs; Berkshire and the Home Countries, often hand-made brickwork with solid brick walls and some half-timbering with tiled roofs; Edinburgh with its gaunt, begrimed stone buildings; Aberdeen with its durable granite; North Wales with its slate; Essex and Suffolk with timber framing and tarred weatherboarded walls – the list is inexhaustible.

Once again it must be stressed that the surveyor should be very aware of traditional materials and traditional methods of construction for that particular locality in which he is working. This awareness can only be obtained through experience. It is the surveyor's task to study historical building methods perhaps more than new forms of design and construction since more often than not it is the older properties which will demand his attention and on which most of his time will be spent.

Referring to the use of traditional materials, as stone is one of the most widespread in use for the older buildings, a few notes on the weathering properties of the commoner building stones may be useful to the surveyor:

1. *Bath stone:* A variety of types from cream to light cream with a more open texture than Portland stone and of poor weathering qualities in industrial atmospheres. This stone is mined, it is very easy to work, and is capable of being easily cleaned.
2. *Beer stone:* Fine-grained white limestone from the East Devon coast, not very durable for external use but found in many South West churches and, in particular, Exeter Cathedral. The quarries have had to be re-opened to replace the badly weathered stone of that particular structure.
3. *Clipsham stone:* Coarse-grained light brown limestone from near Rutland used in Houses of Parliament restoration and in the Oxford area.
4. *Darley Dale:* Light grey to buff fine-grained Derbyshire sandstone with excellent weathering qualities but very difficult to clean.
5. *Doulting:* A coarse, crystalline Somerset limestone coloured grey to light brown and containing shell fragments. Unsuitable for city or industrial atmospheres or for fine detailing as then likely to disintegrate on arrises.
6. *Forest of Dean:* A blue/green/grey Gloucester sandstone with fine grain, used for monumental work and paving as well as for facings and structural work as it weathers very well.
7. *Ham Hill:* A beautifully-coloured rich yellowy-brown Somerset limestone containing shells but soft and unsuitable for use in built-up areas. Cleans well.
8. *Hopton Wood:* Creamy Derbyshire limestone containing fossils and similar to marble when polished but unsuitable for external use in city atmospheres.
9. *Kentish rag:* Grey-white limestone much used in London and the South East. Cleans well and used largely for rubble walling.

10. *Portland stone:* A Dorset limestone, almost white and probably the best limestone for use in city atmospheres as it weathers, giving a 'light and shade' effect. Encountered in large buildings of many towns and cities and particularly suitable for classical detailing.
11. *Wootton (or Runcorn):* Found in Cheshire and South West Lancashire, and generally a durable, fine-grained red sandstone much used in external works but weathers badly in marine atmospheres.
12. *York stone:* A brown Yorkshire sandstone with a fine grain and mainly used for pavings, sills and dressings externally. Very durable but not easy to clean.
13. *Granites:* These vary in texture and quality, the best probably being found around Aberdeen and on Dartmoor (Devon). Very hard and durable, it varies from pink through grey to brown. Used in engineering works as well as building. Beware of the brown granite from West Cornwall, however. This type can be porous and have poor weathering qualities as it is in the first stages of deterioration into china clay.
14. *Slates:* Obtained from various parts of the country, a variety of slates used in thin-bedded courses and varying in durability. Much used in the Lake District, Isle of Man, Wales, central Cornwall and the borders of Dartmoor. Should be laid with horizontal joints sloping *outwards* to prevent damp penetration.

Buildings of special interest

Many of our older buildings are of special architectural or historic interest, and this fact must be borne in mind when any old building comes under survey since special controls and restrictions could be involved. Grants towards the costs of repairs and maintenance may be obtainable but in many cases no alterations of any kind may be carried out without planning consent.

The Secretary of State for the Environment is required under Section 54 of the Town and Country Planning Act 1971 to compile or approve lists of buildings of special architectural or historic interest, to notify this fact to owners and occupiers concerned, and to register the building as such in Part 10 of the local land charges register. When this has been done and a building is included in a list, then 'listed building consent' must be obtained under Part IV of the Town and Country Planning Act 1971 before any person may demolish or carry works for the alteration or extension of that building 'in any manner which would affect its character as a building of special architectural or historic interest'. Section 55(6) of the 1971 Act dispenses with this requirement in the case of works urgently necessary for reasons of safety or health, but notice must still be given to the Royal Commission on Historical Monuments (England and Wales) in such cases if demolition is involved and to the local authority in any event.

Section 58 of the 1971 Act states that a planning authority may serve a building preservation notice which then 'lists' that building for six months together with all the ancillary restrictions involved.

Section 277 of the 1971 Act deals with 'conservation areas' and the local authority must, under the terms of this Section, give public notice of any planning application affecting property within that conservation area in addition to which any demolitions require consent under the Town and Country Amenities Act 1974, Section 1.

Local authorities may serve the owner of a listed building with a repairs notice to ensure the preservation of that building, or may even acquire such buildings compulsorily, compensation being reduced if deliberate dereliction of the building has been allowed by that owner. In some cases, a local authority may carry out repairs and then recover the cost and it is at this stage that the precise condition of a listed building becomes vitally important in the surveyor's report to his client.

In the case of a listed building, advice should also be included in the surveyor's report on the possibility of obtaining a repair grant. Such a grant would be in addition to any other normal house renovation grant to which reference will shortly be made. The Society for the Protection of Ancient Buildings would give further advice on this or any other related subject.

A local authority may contribute towards the repair and maintenance costs of a listed building under the provisions of the Local Authorities (Historic Buildings) Act 1962, including the upkeep costs of garden and outbuildings in some cases. In addition to a grant, interest-free loans are sometimes given, but a grant may be recovered by the local authority if the building is sold within three years.

Outstanding historic buildings may receive repair grants under the terms of the Historic Buildings and Ancient Monuments Act 1953 and this should be mentioned in a survey report if thought to apply and subsequently confirmed.

It must be remembered, however, that listed building consent does not convey immunity from building regulation and planning permissions which must be sought where applicable.

Grants

Continuing on the subject of grants, a prospective purchaser may be interested in obtaining grant assistance from central or local government, or from other agencies, in order to carry out essential repair work to install basic amenities in an old building. Legislation on this topic changes from time to time and when grants are discretionary the local authority's grant policy may be determined by the availability of funds.

If advice is to be offered to a client it is *essential* that the surveyor is sure of the facts. It would be unfortunate if a client were told that a grant would be available, went ahead and bought, and then found that the works involved were not eligible for grant assistance.

In the past improvement grants, intermediate grants, repairs grants and special grants were variously available from local authorities together with home insulation grants for lagging plumbing and ceilings in roof spaces. These grants were either mandatory or discretionary, but they were not means tested.

The Local Government and Housing Act 1989 has made major changes to the system by which local authority grants are made to home owners, landlords and tenants for the repair and improvement of their houses. From July 1990 eligibility is to be determined by reference to a standard of fitness and by a means test of the applicant's resources. There are now five types of grant.

The *Renovation Grant* replaces the improvement, repair and intermediate grants; it is available by right for works to bring a dwelling up to the required fitness standard but the amount payable will depend upon the applicant's resources determined in relation to the Housing Benefit system. The standard of fitness by which a property is assessed as being fit for habitation has been broadened to include artificial as well as natural lighting, heating, a suitably located w.c., bath or shower, wash basin and sink with hot and cold water.

The *Disabled Facilities Grant* is also means tested and can be used for adapting a dwelling to suit an occupier's disability. A *Minor Works Grant* of up to £1 000 is available for elderly owners or tenants, this being discretionary and only available to people receiving income support, family credit, community charge rebate or housing benefit. Applicants may make successive applications subject to a maximum of £3 000 over a three year period.

Grants are also available for houses in *multiple occupation* and for repairs to the common parts of buildings containing flats. These grants will be available to landlords subject to a test of resources.

This Act also changed the old system whereby areas could be designated Housing Action Areas and General Improvement Areas. Now there is a power to declare *Renewal Areas*. It is envisaged that these will be fewer in number, but larger, than Housing Action Areas or General Improvement Areas and within these new areas a mixture of building improvement, clearance and environmental improvement works will be encouraged.

The overall thrust of this legislation is intended to concentrate grant assistance on properties in the poorest condition and to make it easier for poorer households to obtain grants. Grants can, in theory, cover 100% of the cost of works if the applicant's income and resources are low enough.

The main beneficiaries of the new system will be the elderly and infirm. The test of resources favours those without mortgages since mortgage payments are not taken into account when assessing income and expenditure. The system will be unlikely to help purchasers of old properties in poor order who have a large mortgage and who might, in the past, have used grant assistance to pay for the necessary works.

Principal defects

Finally, probably the most damaging features in old buildings for which the surveyor must keep a keen watch are principally:

1. *Brickwork:* Ground movement due to clay, made ground or changes in groundwater levels.
   Thermal movement due to fluctuations in temperature and differential expansion between varying materials.
   Drying shrinkage where calcium silicate (sand/lime) bricks have been used.
   Deterioration from intense heat and sudden cooling as in the case of a fire.
   Roof spread due to incorrectly-designed pitched roofs.
   Sulphate attack from soil or bricks themselves.
   Corrosion of embedded metal, such as ties, beams and reinforcement, causing expansion.
   Defective mortar which is too weak in mix, or use of imperfectly-slaked lime.
   Frost damage in areas of severe exposure such as parapets, retaining walls and brickwork below damp-proof course.
   Salts in brickwork causing efflorescence and subsequent spalling and disintegration after crystallisation.

2. *Structural timbers:* The various types of fungal decay and insect infestation have been fully described beforehand, but very old buildings containing structural elm, chestnut, walnut and oak are particularly prone to infestation by death watch beetle (*xestobium rufovillosum*). Very old structural timbers, most probably having been exposed to previous fungal attack, are the most likely to fall victim to death watch beetle infestation which can lead to severe structural weakness and eventual collapse. Organic, solvent-based liquid treatments containing insecticide have in the past not proved to be fully effective, and on hardwoods a spray application is normally insufficient on such non-absorbent surfaces. Insecticide smokes are generally accepted as the most effective method of eradication and must be used during beetle emergence in late March and early April.

It must be borne in mind that many of the visible exit holes in affected timbers may be hundreds of years old and the only certain method of assessing the position and extent of the infestation is to make a beetle count over one emergence period as beetles fall to the floor on emergence, usually under the area of infestation.

The beetle life-cycle is usually four to six years but can be up to 15 years and consequently perhaps up to ten annual treatments with Lindane or Dieldrin-based smoke may be necessary at the emergence periods.

3. *Stonework:* Main defects in structural stonework may be caused by the following:

Fractures caused by rusting iron cramps and ties (bronze and copper should be used in preference).

Frost action causing cracking and spalling by freezing and expansion.

Figure 15.4 Death watch beetle, common furniture beetle and wood boring weevil have all attacked this old damp hardwood

Atmospheric pollution, particularly that causing decay in limestones due to acid action and destruction of the protective skin which forms over a stone surface.

Incompatibility of stones, for example, association and juxtaposition of **limestone with sandstone in polluted atmospheres.**

Vents, shakes and soft beds causing inherent defects in the stone which affect weathering qualities.

Incorrect bedding where structural stone has not been laid on its natural bed.

Careless quarrying involving excessive blasting and incorrect use of tools.

Crystallisation of salts formed when wet stone dries out and consequent damage by expansion.

Seasoning – not allowing the moist 'quarry sap' to dry out before using the stone.

Living agents of decay such as ivy (which secretes acid), ferns, wallflowers, buddleias but *not* lichens and moss which are not generally harmful.

Damage caused by fire, particularly in the case of the calcining of limestones, a condition that was observed during the 1939–1945 air raids on the UK.

Defective maintenance of rainwater pipes, gutters, copings and pointing causing damp penetration and reducing the life of the stone.

4. *Roofs*: 'Nail-sickness' to slate roofs causing slates to slip where nails have rusted through, often evidenced by extensive clipping back of slates with lead, zinc or copper clips known as 'tingles'. If slates are sound they may be lifted and renailed on new battens and sarking felt since good Victorian slates have a life span well in excess of that of Victorian slater's nails which were not galvanised.

Lamination and softening of clay tiles or stone roofing slabs caused by frost damage and efflorescence.

Defective roof detailing especially around stacks, behind parapets and at other vulnerable points.

Leaking flat roofs, especially balconies subject to foot traffic; watch out for lead, copper and especially zinc which has become thin and brittle from the effects of acid attack especially in industrial environments – also such roofs which may have been recovered with asphalt or mineral felt with only a short life.

Loose parapet coping stones, chimney pots and other items which could cause physical danger to passers-by particularly on commercial buildings abutting the public highway; also rusted, rotted or badly designed balustrades and safety rails to high level fire-escapes and balconies.

Wet rot and beetle attack to exposed rafter ends, fascia and soffit boards and gable-end timbers; a useful tip when inspecting such areas from ground level against the sun is to use a mirror to reflect sunlight back under soffits.

Chimney flues closed off with no ventilation at the base causing condensation inside flues – the accumulated tar deposits from past fires which are absorbed by soft parging in the flues will often cause severe brown staining to interior plasterwork if conditions within the flue are damp, and similar staining will be caused if gas or oil-fired boilers are connected to old flues without first relining. Where fireplaces have been bricked-up this can be confirmed by counting the chimney pots and relating this to the number of fireplaces or vents provided inside; clients should be advised to make up any deficiency by inserting vents top and bottom into any unvented redundant flues and capping off the flue terminal to prevent rain penetration.

From the text of this chapter, the surveyor may conclude that building surveys of old buildings – particularly churches, country houses and public buildings – is a specialist field, but if due consideration is given to the types of construction and materials used, as described in this chapter, then a logical appraisal of such a building on these lines will bring the desired results.

Chapter 16

Reports on leasehold properties

The managing director read the lease, and could not understand Clause XIV. Then he read the accompanying solicitor's letter regarding Clause XIV and could not understand that either. So he asked for a letter from them clarifying their first letter and could not understand that....
Extract from client's letter of instruction

If a client is contemplating taking either a new lease or an assignment of the existing lease of a property, certain special considerations apply to the advice to be given, taking into consideration the particular position of the lessee (the tenant) under the law of dilapidations.

Leases of houses

A number of different possibilities may arise. In Manchester, Cardiff and certain other localities, dwellinghouses have been sold traditionally on a leasehold basis, generally for a 99-year or 999-year term. The reasons for this are essentially historical and arose because the landholdings in these areas were in the hands of certain families. Initially, the idea was good since after 99 years and the expiration of the leases, the general redevelopment of an area might have been contemplated and the ground landlords would have retained some control over the properties for the benefit of all concerned. Unfortunately, the results in practice were that leasehold dwellinghouses tended to fall into an increasing state of disrepair as leases began to run out because the lessees were unable or unwilling to improve them.

As a result of pressures for reform of the leasehold system, the 1967 Leasehold Reform Act was passed which gave leaseholders with certain types of long leases of houses (but not flats) the right to buy the freehold, or the right to a specially-extended term at a moderate ground rent subject to proving the required period of owner-occupation and certain other conditions.

Unless the lease term is short (say less than 30 years) it is felt that it would be unusual in practice for the ground landlord of a leasehold house to take very much interest in its condition, and it would be most unlikely that a schedule of dilapidations would be served except in unusual circumstances. In most cases the report on such a leasehold house will follow the same lines as would be the case with a freehold property.

Perhaps the main point which could be made to a client contemplating buying a leasehold house would be to suggest that he might acquire the freehold as well. He could do so by asking the vendor to serve the necessary Notice of Leaseholder's Claim and transfer the property with the benefit of that Notice. If this is not done, the purchaser cannot take advantage of the Act himself until he has first served out the necessary term of occupation. The client should be advised to consult his solicitor on this point. In the long run, a freehold house is a far better investment.

Where the lease of a house is running out and has only a short time to remain, then the necessity for a schedule of dilapidations may well arise. Shorter leases of houses will be encountered in some fashionable London districts, and rights to enfranchise under the Leasehold Reform Act may be different or may not exist, depending upon the size and value of the property concerned. A careful reading of the lease in such circumstances is recommended before giving the appropriate advice.

Leases of flats

Leases of flats and apartments are the rule rather than the exception in England and Wales for the reasons discussed in Chapter 14. An almost infinite variety of forms of such leases will be encountered in practice. However, the surveyor will find that these generally fall into one of two basic types: either there will be a lateral division of responsibility within the structure of the building with each leaseholder responsible for his own part of the structure and having no liability for other parts (mostly encountered in cases of small two- or three-storey buildings in flats, either purpose-built or converted), or the lessee will be responsible for the interior of his own unit and share responsibility for the exterior and structure as a whole with other flat owners. In the latter case, arrangements for the proper maintenance and upkeep of the exterior and structure may simply be on an *ad hoc* basis as the need arises. Alternatively, this may be the responsibility of the freeholder who would then recover costs from the lessees, or commonly, may be undertaken by a management company or residents' association.

It will be seen in such cases that a lessee may have a potential proportional liability for parts of the building, or even parts of other buildings on the estate some distance away from his own property. Particularly onerous repair liabilities may arise in the case of flats in older mansion blocks in the inner cities.

Common practice is for such a mansion block to be acquired by a speculator, and for individual flats to be sold off on new long leases with a ground rent and service charge for the maintenance and repair of the exterior, roof and common parts. Initially, in order to sell the leases, the freeholder will wish to keep the service charges low, perhaps artificially low. Once the units are all sold, however, there is no such need and when the true state of repair of the structure begins to become known, large increases in the service or supplementary charges may be levied to cover the cost of repairs which should have been dealt with at the outset during refurbishment and prior to sale.

Before undertaking an inspection of a flat in such a mansion block, the surveyor should be quite sure that his survey will be sufficient to gauge whether or not any exceptional service charge increases may be due. If this is to be done, it will be clear that the inspection cannot be confined merely to the one specific flat. The significance of this is discussed more fully in Chapter 14.

In all cases of surveys of leasehold residential property, the RICS Code of Practice advises that the surveyor should read the lease before undertaking his inspection. He must endeavour to establish the extent and nature of the repairing liabilities under that lease, the client's potential responsibility for executing repairs and also his liability to pay for repairs executed by others. If the lease is not available, the surveyor must set out clearly the limitations inherent in his advice.

If no copy lease is available, the client's solicitor should be approached and asked to confirm the nature of the repairing convenant which may be assumed. The surveyor may then, on receipt of such instructions in writing from his client's solicitor, refer to this in compiling his report.

The Code of Practice suggests that when taking instructions, in addition to the procedures recommended for freehold property, the surveyor should insist either on having a copy of the lease or on having advice on the nature of repairing liabilities. He should then confirm instructions in the manner recommended for freehold property, indicating also the extent to which he proposes to inspect other parts of the property or estate in order to advise on unacceptable or unusually heavy repairing liabilities outside the demised premises.

Leases of commercial premises

Dealing with leases of commercial premises, the recommendations for procedure could be similar to those for leasehold residential property in that it is highly desirable for the surveyor to read the lease before he undertakes his inspection and to have it available for references as he compiles his report. He may then consider:

(a) The repairs or other works which are required for the purposes of the client's business occupation and necessary to comply with the various local and national regulations, and
(b) the future repairs or other works which may be required in order to comply with the repairing convenants of the lease.

In the case of (a) above it is necessary to ascertain the nature of the client's business, how he proposes to use the premises and the staff which will be employed. In respect of (b), the surveyor will have regard to what the lease says and the current relevance of the law of dilapidations (see Chapter 20).

Section 18(1) of the Landlord and Tenant Act 1927 sets out certain principles in the assessment of damages for breaches of repairing covenants. It provides that the measure of damage for breach of a covenant to keep premises in repair, or leave premises in good repair when quitting, cannot exceed the diminution in the value of the landlord's reversion occasioned by these breaches. A limit is therefore imposed on the amount of damages which may be recovered by a landlord. These are assessed by valuing the premises as they are and as they would be if the covenants had been complied with, the measure of damage being the difference between these two figures. The cost of undertaking remedial works may be the same thing as the difference in value, on the assumption that a purchaser of the lessor's interest would reduce any price offered by an amount equal to the cost of remedying those defects. The cost of undertaking works may, however, be an amount quite different from the measure of damages in some cases. It is possible for there to be considerable breaches of repairing covenants by a lessee and for the landlord to have no right to compensation at all (for example if the landlord is proposing to redevelop the property in such a way that condition is immaterial to market value).

The Courts have held that the cost of undertaking works is *prima facie* evidence of the diminution in the value of the reversion, which assumption can be rebutted by evidence that damage to the lessor's interests may be better assessed by some other means, such as 'before' and 'after' valuations. In *Drummond* v *S. & U. Stores Ltd.* (1980) 258 EG 1293, *Estates Gazette*, 27 June 1981, p 1293) it was held that in the absence of clear evidence as to the damage to the reversion, the cost of undertaking necessary works was the broad measure of the diminution of value and the landlords were awarded the bulk of the cost of repairs. The landlords in this case were also awarded three months' loss of rent as part of the damages, being the estimated time reasonably necessary to leave the premises empty while the works were done, and also, as the landlord in that case was not registered for VAT, that VAT on the cost of repairs would also be recoverable.

The surveyor advising a client on a commercial lease should have uppermost in his mind the possibility of either an interim or final Schedule of Dilapidations being served on his client and should warn the client about all matters which could give rise to the service of such a schedule or a claim for diminution of value. Substantial sums may often be involved in such cases. One means of protecting a client at the commencement of a lease is to arrange for the lease to state that the lessee will not have to hand the demised premises back on expiration of the term in any better condition than they were in at the commencement as evidenced by a Schedule of Condition. An agreed Schedule of Condition is then prepared by the lessee's and lessess's surveyors, signed and attached to both lease and counterpart lease.

During the course of taking instructions to inspect a leasehold property, the surveyor should ascertain whether or not a Schedule of Condition exists which indicates the condition of the premises as at the beginning of the term. If such a Schedule exists, then it would be advantageous for the surveyor to have a copy and to take this with him when he carries out his inspection. The condition of the property as described in the Schedule of Condition may then be compared with the condition found on inspection.

If a building is to be inspected when a client proposes to take a new lease, and if that building is old or in poor repair, then the surveyor may recommend that a Schedule of Condition be prepared and agreed with the prospective lessor prior to exchange of the lease and counterpart. This may be especially important where a modern form of full repairing lease is being granted on a building which is old and which may suffer from signs of structural distress, past foundation movements, **faulty roof construction and other matters potentially expensive to rectify. In** general, the lessee's liability is for repair, and where a defect can be repaired then the liability is not to renew or rebuild, nor will there be any liability to make a building better than it was previously. In practice, however, there is considerable room for argument as to the extent of repairs which may be required.

Perhaps a roof covering is in poor condition and should be stripped off and renewed if it is to be made satisfactory. However, a repair in the form of overhaul and patching of the covering may suffice in the short term. There could be considerable room for disagreement between lessor and lessee as to the extent and nature of the pairs in such a case. A Schedule of Condition detailing the condition of the roof covering at the commencement of the term would provide a standard against which any repairs or renewals may be judged adequate or otherwise.

It may be that a property suffers from dampness. Generally, there is no liability for the lessee to provide damp-proof courses or other works where none previously existed, although in such circumstances he will be liable to deal with the effects of dampness if these have caused deterioration of the interior. If damp-proof courses were in good condition at the commencement of the lease term but failed and became porous during that term, then a liability to renew damp-proof courses could arise. In such a situation it may be difficult to confirm all the facts and if a Schedule of Condition is discovered which clearly shows that the property had a damp problem when the lease began, then this could be of considerable benefit to the lessee's case. Conversely, if a Schedule of Condition indicates that the property was dry and the damp-proof courses were in good condition at the commencement of the term, then this would assist the lessor in requiring remedial works to be undertaken at the lessee's expense, or in obtaining damages for diminution of value.

On occasion, a building may suffer from structural movements of one kind or another. In old buildings such matters rarely arise overnight. More often than not they are part of a continuing process with crack damage, bulging and leaning walls and other manifestations having appeared very gradually over a long period of time. Rarely could a lessee be found liable to undertake a structural repair such as underpinning, but there could arise a liability to provide buttresses to leaning walls or tie bars to stiffen the structure. An orginial Schedule of Condition is helpful in such cases if it accurately describes the structural condition of the building in detail as existing at the commencement of the lease term.

In addition to the repairing covenants in a lease there will also be other covenants, some not directly relevant to the survey but others which may affect the inspection and preparation of the report.

There will generally be a 'user clause' in the lease indicating uses which may be unacceptable and those which may be permitted. For example, a client who wishes to use premises for purposes other than those acceptable in the lease must first obtain the lessor's consent for the change of use. If the lease provides that such consent may not be unreasonably withheld, then the lessor, in determining whether or not to grant consent, is subject to a test of reasonableness which may be challenged. If in the lease there is no provision as to reasonableness the lessor may or may not grant consent for a change of use at his discretion. Clearly, prior to any exchange of contract to acquire a lease, a client must obtain a determination in such cases.

There may be objections to a specific retailing use which conflicts with other retailing uses in adjoining premises which are owned by the same lessor. A lessor who owns a number of retail premises in the same shopping district will wish to ensure a balance of different types of retailing in the area for two reasons, one being the need to present a broad range of retailing outlets to make the centre attractive and profitable, and the other being to limit competition between shcp units selling identical products or services which could otherwise make one, or both, competing units unprofitable.

Objections to use may arise out of the nature of the uses themselves since dust, noise, smells or other problems could arise to the annoyance of the lessor or other adjoining lessees. Certain specific trades will often be particularly referred to in a lease as being unacceptable. These will often include panel-beating, fish-and-chip shops, heavy industry in a light industrial area, or trades such as glue factories or fertiliser storage which could cause offence.

There will generally be some mention of 'assignment' in the lease. A lease with an absolute bar on assignment cannot be sold and is of value only to the lessee in possession as it has no value as a mortgage security. Where assignment is permitted with consent, the law holds that such consent may not be unreasonably withheld, even if the lease omits to state this, so that assignment to a lessee of comparable standing will normally be permitted subject to the lessor providing a Licence to Assign, generally on payment of a small fee to his solicitor.

Covenants in the lease barring structural alterations to the buildings demised are generally found. Such covenants will generally refer to the need for specific permission to be given by the lessor for any alterations to the premises, but an absolute bar on alterations may be provided as an alternative.

A lessee who wishes to undertake alterations to business premises which will constitute improvements and which may increase the value of the premises may, if he wishes, serve notices under the 1927 Landlord and Tenant Act. This is often overlooked, with the consequence that a lessee loses any rights to compensation for improvements or additions to the landlord's property when he vacates. Advice about 1927 Landlord and Tenant Act procedures needs to be given at the outset before improvements are made and should be dealt with at the planning stage.

Proposed changes in leasehold law

At the present time proposals are being formulated to allow the owners of flats and maisonettes to purchase their homes with commonhold title whilst at the same time allowing enforceable arrangements to arise to deal with the maintenance and upkeep of common parts and the payment for shared services. Such reforms have been long overdue and pressure is now growing from the owners of leases granted some years ago where the shortening lease term is having a serious effect on the resale value of the property. It seems likely that the law on this matter in England and Wales will be changed and that flats will generally be sold on a freehold basis in the not too distant future. Inevitably this must mean that the rights to buy freeholds at present limited to owners of houses will be extended to the owners of flats.

Consideration is also being given to the problems associated with the management of blocks of flats, which is a matter inevitably tied up with the existing leasehold arrangements which have been found to be unsatisfactory in many cases. Again reform of the law to provide access to special courts to settle disputes and order works to be carried out may be the answer as serious difficulties are arising at the present time with the management of larger blocks of flats, especially in the cities where many such buildings are old and in poor repair.

Existing arrangements permitting leaseholders of flats to club together and buy the Freehold compulsorily, or extend their leases, have been found to be complex and difficult to apply in practice.

Chapter 17

Reports for prospective mortgagees

If language is not correct, then what is said is not what is meant; if what is said is not what is meant, then what ought to be done remains undone
Confucius

Mortgagees are primarily concerned with ensuring that the security they take for the purpose of the mortgage will be readily re-saleable at a price which will allow for the repayment of any monies due and cover the incidental costs involved, including the costs of sale.

Institutional mortgagees will also wish to ensure that their actions are seen to be in the public interest and they will want to foster the goodwill of their mortgagors and consumers generally. With this in mind, the main building societies and banks now release copies of their surveyors' reports to the mortgagors.

It should be mentioned at this stage that the mortagor grants the legal mortgage in return for the loan and is the borrower (and purchaser). The mortgagee takes the legal mortgage, grants the loan and is the lender. It is not uncommon for the terms to be wrongly transposed and it is important to use the correct nomenclature.

Reports for prospective mortgagees may take the form of pro-formas or be the traditional typed reports on the surveyor's own paper. Such reports may or may not include a full building survey but they will inevitably include some comments on the condition of the structure since it is not possible to value a property without first forming an assessment of its general condition.

If the property is in poor repair the surveyor will be expected to give the mortgagees some guidance on the significance of this and the conditions which should be applied to any loan. Such conditions may take the form of retentions from the advance money pending completion of essential works and a further surveyor's inspection, or may simply require an undertaking from the mortgagor to complete the works within a specific period of time.

The advice to be given on repairs required will be dependent to some extent upon the amount of advance involved. Generally, low-percentage advances involve low risk levels for the mortgagees and in many cases the requirement for repairs will not be critical. In such circumstances, a written undertaking from a mortgagor of good repute to complete the works within a specified time would seem adequate.

High-percentage advances involve higher risk levels to the mortgagees, and experience tends to indicate that the risk of default by a mortgagor tends to be higher in the earlier years of the mortgage term when the outstanding balance of capital is at its highest. In such circumstances, it may be necessary for the mortgagees to retain part of the agreed advance and to release this against certification from the surveyor when essential repairs have been completed. Building Societies are especially vulnerable in this respect because they provide mortgage advances to house buyers in circumstances where the percentage advances are high (sometimes as much as 100%) and the security is often an older building in poor repair.

The mortgage transaction is achieved by means of a mortgage deed in which the mortgagor agrees to terms controlling the repayment of capital and interest and may also agree to certain requirements for the maintenance and repair of the property used as security over the period of the loan, also the insurance of the building against specified perils. The mortgagor retains the right to recover the property title free from the encumbrance of the mortgage, on payment of the amount due plus costs and penalties if the loan is redeemed early. The right of the mortgagor to redeem the loan is known as his 'equity of redemption'.

Virtually all mortgages the surveyor has to deal with will be mortgages granted by a charge on the property expressed by way of a legal mortgage. Since 1925, this has been the principal method of lending money to a property owner against the property security and is the only legal form of mortgage apart from a much less usual procedure for the grant of a lease to the mortgagee with provision for cessation on redemption. Occasionally, an equitable mortgage is encountered which may be a contract to enter into a legal mortgage if required, or a verbal agreement accompanied by the deposit of the title deeds or lease of the property with the mortgagee. This latter procedure is occasionally adopted by bank managers when granting small loans to achieve security without the need for a full legal mortgage. It does not confer a legal interest in the property and is not suitable for larger-percentage long-term advances, such advances invariably taking the form of a legal charge.

The security for the loan lies in the ability of the mortgagee to take possession of the property and sell it to recover the loan together with incidental costs. There are a number of ways in which a mortgagee may take steps to protect his position if the repayments under the mortgage deed are not being met. No mortgagee likes to take such steps and institutional mortgagees such as banks and building societies will go to considerable lengths to avoid such procedures. They may give the borrower an extended time to pay or a capital repayment 'holiday' to see him through a difficult time. Only as a last resort will they take steps to obtain possession of the property, especially in the case of residential owner-occupied homes, although the number of homes taken into possession has increased in recent years. Local authorities tend to be lenders of last resort in some sections of the residential property market and the incidence of default by local authority borrowers is rather higher, as is the incidence of default by borrowers on the security of property investments and commercial premises.

Generally, the procedure adopted when all else fails is for the mortgagee to apply to the Courts to take possession. A mortgagee in possession will normally be able to require the borrower and his family to vacate if the property is an owner-occupied house and will also be able to obtain vacant possession in other cases where the borrower, or borrowing company, is also the occupier. The property would then be sold in the open market for the best price obtainable at the time. Any surplus left over after the loan, incidental expenses and accumulated interest have been paid must be paid to the mortgagor. If the property is subject to tenancies and the tenants are protected by virtue of the Rent Acts, Landlord and Tenant Acts or Agricultural Holdings Acts, then the mortgagee takes legal possession subject to the subsisting tenancies. The existence of tenancies may therefore substantially reduce the re-sale value of the mortgage security. When a surveyor inspects the security he should look out for any signs of tenancies about which the mortgagee may be unaware and suggest, if appropriate, that the existence or otherwise of tenancies be confirmed.

An alternative procedure would be for the mortgagee to apply to the Courts for a foreclosure order which would have the effect of extinguishing the mortgagor's equity of redemption and transferring the property to the mortgagee. For technical reasons this procedure is less frequently used than the procedure for the mortgagee entering into possession, but in certain situations foreclosure can have advantages.

Another alternative procedure is for the mortgagee to take possession but with a view to managing the property, collecting the income and paying any outgoings. The net income could then be used to reduce the mortgage debt and pay the interest charges with any arrears. In modern times, such a procedure would not normally be used for vacant property unless the premises concerned proved difficult to sell in the open market but were potentially lettable. These circumstances could arise with certain types of commercial or industrial premises, premises affected by planning blight or having other problems which would be a disadvantage to the potential purchaser but not to a potential tenant.

It may be that a mortgagee in possession will take over premises which are tenanted, either because the premises were already let and the loan was granted on the security as an investment, or because the tenancies have been created after the grant of the loan. Generally the mortgage deed will provide that the mortgagee's consent is required for any letting of the property which secures the loan, but lettings may be undertaken unlawfully without the mortgagee's consent. In such situations it is more likely that the mortgagee in possession will not seek to sell the property immediately since the value may be very low either because of the existence of tenancies or the climate in the property investment market may be unfavourable. It may then be advisable for the mortgagee in possession to manage the property and use the income to pay off the debt as well as meeting the interest due.

In valuing a property for mortgage purposes the surveyor will then need to consider two matters. First, the re-sale value of the property should be sufficient to cover the amount of the loan plus incidental costs. If this is not so, then the mortgagee will suffer a loss. In certain circumstances where the loss suffered is due to the surveyor's valuation being too high, the surveyor himself may be held to be negligent. Secondly, in the case of premises which are tenanted, or are to be tenanted, the surveyor may need to consider whether or not the net rental income from the property will be adequate to cover the repayments. Clearly it would be unwise in most cases for a mortgagee to advance money on the security of a property investment in circumstances where the income from the investment is less than the mortgage repayments.

Certain types of property represent good security for mortgage advances. Owner-occupied houses with vacant possession and in good condition are good security since there is a broad market for such houses in most localities in all save the most dire economic circumstances. Residential investments are a less satisfactory security especially as the net income will rarely be sufficient to cover the repayments on a substantial percentage advance. In modern times, much of the value of such investments lies in their potential for future vacant possession and such values are prone to the effects of general economic circumstances, interest rate changes and changes in legislation.

Commercial, industrial and other types of building will have market values and potential rental incomes which will vary greatly in different localities and in different economic circumstances. The more specialised or remote such a property is, the greater will be the degree of risk of a fall in value. Some types of older industrial building may become unsaleable or unlettable in time if a suitable alternative use cannot be found. The implications of such a situation for the mortgagee will clearly be very serious.

In making a valuation of a property for mortgage purposes, the surveyor should follow the ordinary principles of good practice. Since this book is not intended to cover the principles of valuation, it is not proposed to discuss techniques. However, there are certain matters which will have to be confirmed during the course of the survey of a building for a prospective mortgagee because the information is needed in order to prepare a valuation.

The first of these matters to be ascertained is the condition of the building. Although the survey may not be a full building survey and the inspection may be abbreviated, nevertheless it is not felt that any building can be valued unless the surveyor has inspected it inside and out, however superficially, and formed some view as to its condition. Clearly the possibility of serious defects in a building, especially an older building, has to be considered and if such defects are present they will need to be taken into account in formulating a valuation. If such defects are suspected but cannot be confirmed in the time available or without special exposure, then a provisional or conditional valuation may have to be issued and the prospective mortgagee will then have to consider whether to pursue the matter any further in the light of the percentage loan applied for.

It is usual for the valuation report in many cases to contain caveats or conditions having regard to the possibility of concealed structural defects, high-alumina cement concrete in widely-spanned commercial structures, and similar hidden problems. In many such cases the mortgagee will require the mortgagor to obtain a technical assessment of possible structural problems of this type as a pre-condition of any advance. This is a useful way of safeguarding the mortgagee's and mortgagor's positions since any such matters which come to light will be of interest to them and may be crucial.

Measured surveys

The second matter to be assessed is the physical size and shape of the building so that it may fairly be compared with other similar buildings for valuation purposes. This will generally require a measured survey to confirm the floor areas, clear spans, ceiling heights and usable frontages. In 1979 the Royal Institution of Chartered Surveyors and the Incorporated Society of Valuers & Auctioneers jointly published their Code of Measuring Practice. It is recommended that the practices set out in this Code be adopted wherever possible. The Code recommends that gross internal area (GIA) as defined should be used for industrial and warehouse space, although agency practice in many areas is to use gross external area (GEA) when dealing with new industrial and warehousing buildings. The Code also recommends measuring procedures for shops and residential buildings.

The purpose of an accurate measured survey is to enable the surveyor to retain in his records details of properties valued so that details of similar properties may be compared on a like-for-like basis. As long as the basis of measurement is the same, the basis of comparison will be valid and the statistical records kept for valuation purposes will be correct. With industrial and warehouse space, I would commend the use of GIA as being the fairest means of comparison, especially if older properties are being compared with more modern buildings. The usable floor space in an older building with thick walls and other obstructions may be proportionally less than the usable space in a modern structure with clear spans and thin walls. A comparison of GEA in such circumstances may be misleading.

In addition to the two foregoing important matters which have to be resolved when surveying a building for a prospective mortgagee, there are also a number of other incidentals arising during the survey which may be significant. The surveyor may see evidence of tenancies, rights of way, shared access or other complications which could be significant to the mortgagee and which should be mentioned in the report. The mortgagee may also require a pro-forma report to be completed, in which case the surveyor will need to check the questions in that report and confirm that they can be answered fully from the notes he has taken on return to his office.

The mortgagee may require a plan to be prepared showing the curtilage of the property, main plot dimensions and other means of identification such as the distance from the nearest road junction and the north point. The purpose of an identification plan is to compare the property inspected with the title plan to ensure, among other things, that the surveyor has actually inspected the correct building (the possibility of the wrong house being inspected is not as far-fetched as it may seem, especially on a new estate) and that his valuation does not include any property or land which is not part of the holding.

Having completed the survey of the building for the prospective mortgagee, the surveyor will then have to compile his report. The report may be a full building survey, or something rather less, on the surveyor's own paper; alternatively it may be completed using a pro-forma report form. In addition to the necessary description and the valuation for mortgage purposes, the surveyor may need to advise the mortgagee on questions of lending policy, either in respect of a specific property or in a general sense.

General lending policy

General lending policy is concerned with the nature of the properties upon which the prospective mortgagee may feel able to lend and the duration of the loans; the terms controlling the frequency of capital repayment and the rate of interest; and the percentage advance considered suitable in the circumstances.

A trustee, clearing bank or building society lending at relatively low interest rates over long periods will expect near-absolute security for loans. They will, quite reasonably, expect that in the event of a mortgagor's default, they will suffer no loss and the proceeds of sale will cover not only the original advance but also some margin for accumulated interest and costs. Traditionally, an advance of two-thirds or three-quarters of valuation would normally be the maximum loan in respect of good, readily saleable security. Something rather less than this, perhaps 50% or less, may be considered suitable in respect of poorer security. The margin is intended to cover not only accumulated interest and costs in such circumstances, but also the possible vagaries of the market for certain types of property. If the market for a property is doubtful then it may be that no advance at all can be contemplated by an institution at low interest rates and the prospective mortgagor may have to produce some other security. Alternatively, he made approach a lender prepared to accept a greater risk which will undoubtedly involve a higher interest rate on the loan.

During the post-War period, the building society movement has tended to advance rather higher percentages on owner-occupied dwelling-houses with loans of up to 90% or 100%. To enable them to do this, they will normally require some additional collateral such as a mortgage guarantee from central or local government, from an insurance company or a charge on another property or on a life insurance policy.

A surveyor may be called upon to advise a prospective mortgagee on loans being made at higher than normal interest rates on the security of properties which may be declined for this purpose by clearing bank or building society. In this situation it is as well that the lending policy of the clients be clarified at the outset and that the surveyor is able to exercise his judgment in the light of that policy so as on the one hand to protect his client's interests and on the other hand to allow sufficient flexibility for a suitable agreement to be reached.

Mortgagees who advance money at higher than normal interest rates will normally have to accept that the quality of the security offered or the status of the borrower is lower than may be ideal. The higher interest rates which they may charge are intended to reflect this additional risk. Such lenders may be prepared to advance money on the security of residential investment property where the net income is lower than the mortgage repayments, relying upon the borrower's ability to finance the balance from some other source. They may be prepared to lend money on the security of vacant sites, or sites with buildings which are destined for redevelopment where there may be little or no income generated from the property, relying upon the borrower's ability to redeem the advance when redevelopment has been undertaken. They may be prepared to advance money on the security of unusual buildings which are in secondary or remote locations or buildings suffering from defects of one kind or another.

In carrying out a survey for such a mortgagee the surveyor cannot necessarily apply all the principles of lending policy which would be applicable to those mortgagees who charge more conventional lower interest rates. Each situation has to be considered on its individual merits, having regard to the fact that if all principles of good lending policy were to be applied then the institution concerned would be unable to compete with other institutions offering terms at lower rates. On the other hand, if an imprudent lending policy is adopted, the end result may be a spate of defaults should there be a downturn in the property market.

Local authority mortgage

Local authorities make advances on the security of residential property and, in most cases, the intention of providing this service is to enable house-purchase to be contemplated by sections of the community who would otherwise be unable to become owner-occupiers. Local authority mortgagors tend to be younger, seeking to buy houses in the cheaper price ranges with fairly large percentage loans.

Whereas a bank or building society may limit advances to 75% or 80%, the local authority may lend up to 100% of valuation, and whereas certain types of building, such as older houses, converted flats and the like, will be rejected by a bank or building society as security, the local authority may be prepared to accept such property as security for an advance. Many surveys of properties for local authorities are carried out by surveyors who are employed for the purpose by the authority as staff surveyors. Occasionally the local authority will employ an outside surveying practice to do this work.

A feature of surveys for local authority mortgages is that the property being inspected is often older and in poor condition. In the larger conurbations, the building may comprise a converted flat or apartment. In some localities the local authority may be the only potential lender in the market so that the saleability of older houses, converted flats and apartments in some districts will depend almost exclusively upon the funds which the local authority has available at the time. Since the money for local authority lending is subject to the effects of national economic policy, there will be times when loans are granted fairly freely and other times when funds are 'tight' and the availability of council mortgages may dry up altogether.

In carrying out a survey and preparing a report on an older house in poor condition, or a converted property, the surveyor should consider whether it would be reasonable to impose conditions which could make the property more saleable in the future by making it potentially mortgageable to a lender other than a local authority. Converted flats, for example, may lack adequate soundproofing or fire-resistance between units but if these matters are dealt with, perhaps at fairly modest expense, the converted unit may be more mortgageable and therefore more saleable in the future, especially in times when council loans are difficult to obtain. At the time of writing very little local authority lending is being undertaken and their role in this respect has been largely taken over by the building societies.

Leasehold interests

The property being offered as security for an advance may be leasehold. Residential property held on a long lease will not normally be suitable for a long-term advance by way of mortgage if the unexpired term is 30 years or less. If a term of more than 30 years is available, the repayments of capital should be so made that all capital outstanding is progressively reduced until the mortgage is redeemed when the lease still has at least 30 years to run. The reason for this is obvious – a leasehold interest is a wasting asset which will fall in real value during its life. The effects of monetary inflation will mask the fall in value for a time but once the end of the lease is in sight, the fall will be rapid. When the lease expires, the value for mortgage purposes will be nil.

If an advance is made on the security of a short leasehold interest in a residential property, the risk to the mortgagee is high because in the event of default by the mortgagor, there may be considerable delay caused by the legal procedures for obtaining possession. While this delay occurs, the value of the security may be falling. With freehold security, the mortgagee will at least normally know that increasing property values and inflation may be on his side should he find that delay is incurred in proceeding with a remedy for default.

With commercial and industrial premises, a mortgage advance is often made on the security of a lease where the period available before the lease expires, or the remaining period before the next rent review, is quite short. Such leases may have only a very speculative value. In times of economic boom it may be that substantial premiums will be obtainable for such leases, partly reflecting the profit rent (that is the difference between the rent payable and the open market rent which would be obtainable) and partly elements of goodwill, tenants' improvements, fixtures and fittings and temporary scarcity. In times of economic depression, no premium may be obtainable and tenants may be glad to able to vacate by making a straight assignment of the lease without premium. While a mortgage advance may be secured by a legal charge on the lease in such situations, the real security for the advance will not be the value of the lease but will be the ability of the borrowers to repay. If the borrowers are in difficulties, perhaps during a period of economic depression, the fact that a mortgage is secured by the value of the lease may be worth nothing to the mortgagee.

A highly cautious lending policy is therefore recommended where a short leasehold interest is involved. The surveyor should ensure that all the lease covenants have been complied with, especially covenants on usage and repair. He should carefully read the lease to confirm whether or not there are any unusual covenants which could limit the re-sale value of that lease; tight restrictions on uses, in particular, may make a lease unsaleable and therefore potentially unmortgageable. If the building is in poor order, and especially if the condition is such that a Schedule of Dilapidations may be served, the property may need to be put into good repair as a condition of any mortgage advance, and probably as a pre-condition of some, or all, of the advance being released.

When surveying a leasehold property for a prospective mortgagee it is suggested that the surveyor should make a point of reading the lease. The only exception might be in the case of residential property where normal lease terms may be assumed for the purpose of the mortgage valuation and where the mortgagee's solicitors will be able to confirm this and refer back to the surveyor if unusual terms are found to apply. In many cases a building society or bank valuation will have to be undertaken on this basis and a provisional valuation given without the surveyor having seen the lease.

Insurance

A report for a prospective mortgagee will often include advice as to a suitable amount of insurance cover and the mortgage deed will normally provide that the building is to be insured in the joint names of the mortgagee and the mortgagor.

In addition to insurance against fire, there is a wide range of other perils against which cover may be recommended, including impact damage, explosion, storm, tempest, flood, burst pipes and tanks, subsidence, landslip, ground heave, public liability, plate glass, trade fixtures and fittings and loss of rental income or loss of profits. Not all such perils will arise in every case. Subsidence, landslip and ground-heave cover is normally available for residential buildings but not in all other cases. Plate glass and consequential loss cover will apply to certain types of commercial premises.

The insurance should cover all buildings used as security for the advance unless they are not significant to the re-sale value. The sum insured should be adequate to cover the cost of reinstatement in the event of a total loss. If the sum insured is excessive, the mortgagor will be penalised by having to pay excessive premiums. If the sum insured is insufficient then difficulty could arise in the event of a claim not only for a total loss, which rarely occurs, but for a partial loss. The insurers may, in a case of obvious under-insurance, average the claim by paying only in proportion to amount of cover as it relates to the full reinstatement value. For example, if a building is insured for £100 000 and its true rebuilding cost would be £200 000, the insurers may be prepared to pay only 50% of any claims made for insured perils.

In the case of building societies they will generally act as agents for the insurance company who insure all their mortgage securities, often under a single block policy. The surveyor undertaking an inspection of a building for a building society or other institutional lender may need to provide an estimate of the reinstatement cost for insurance purposes and also advise the lenders of any obvious insurance risks which they should then relay to the insurance company.

In low-lying or coastal areas, property could be subject to flooding, and often enquiries have to be made where a building is low-lying in order to ascertain whether or not there is a flood risk. In such cases the insurers may refuse to provide the cover for flood damage or may require additional premiums. Properties with thatched roofs or other types of potentially flammable construction may be subject to premium loading. Buildings located at potentially hazardous points adjoining road junctions or fast sections of highway may be prone to impact damage. Certain types of construction could be prone to storm damage, especially lightweight structures in exposed locations. Subsidence, landslip or ground heave cover may be restricted where a building already shows signs of existing or past structural distress.

It is a rule of insurance law that the parties must disclose any matters which could be relevant to the degree of risk and it is incumbent upon a surveyor who notes any such matter to bring it to the attention of the mortgagee in his report so that the insurers may be suitably advised.

The amount of insurance cover assessed by reference to rebuilding costs current at the time should be periodically reviewed and it is good practice to provide that the sum insured be altered annually in line with changes in building costs. Most insurers of buildings have some procedure for automatic adjustment of the sum insured along these lines. If no such procedure applies, it is suggested that the mortgagees should themselves undertake to review the amount of insurance cover each year to safeguard their position.

The Building Cost Information Service of the Royal Institutional of Chartered Surveyors publish an annual *Guide to House Rebuilding Costs for Insurance Valuation* on behalf of the British Insurance Association. This guide refers to costs of demolishing and clearing away an existing structure and rebuilding it to its original design in modern materials and with modern techniques to the same standard as the original property but to comply with Building Regulations and other statutory requirements. This information is essential where reinstatement costing advice is to be given.

Second and subsequent mortgages

In additional to the provision of first mortgages, it is possible for financial institutions or individuals to lend on the security of second mortgages, or even third or subsequent mortgages.

Clearly, the degree of security for second or subsequent mortgages will be less than that afforded by first mortgages. The second or subsequent mortgagee will be able to recover his amount owing only after the amounts owing to the first mortgagee have been satisfied as far as this is possible out of the proceeds of sale. In effect, the mortgagees have to take their turn in the queue if a mortgagor defaults and the property used as security is sold. It may be that a second or subsequent mortgagee will find it difficult or impossible to recover his debt if the amount owing under the first mortgage is substantial and if there are accumulated costs and interest charges also owing to the first mortgagee.

Having regard to the higher risks attaching to second or subsequent mortgage advances, the interest rates charged are generally higher and the terms for capital repayment often of shorter duration. If a surveyor is asked to survey a building to be used as security for a second or subsequent advance by way of mortgage, then he should find out the amounts of capital outstanding under any first or former mortgages so that he is aware of the amount of the borrower's equity in the property and the extent to which any further mortgage advance will be covered.

If the degree of cover for the second or subsequent mortgage is poor then great caution should be exercised in preparing any valuation and in making any recommendations, bearing in mind the particular need to safeguard the future re-sale value of the security in the light of the state of repair of the premises and all other circumstances.

Mortgagees in possession

Occasionally a surveyor has to return to re-inspect property used as security in circumstances where the borrowers have defaulted, perhaps to advise on the appropriate sale price to quote or to prepare a claim for dilapidations. It is a good idea for surveyors to inspect such property once in a while to remind them of the difficulties which often face mortgagees in such situations. Rarely will the property be in good order; often it will be in very poor condition with serious neglect and perhaps even deliberate damage. The legal process can take considerable time and a borrower in financial difficulties will certainly not be able to afford a high standard of maintenance. In many cases the borrowers may have removed fixtures and fittings, perhaps even stripped the property of anything of value.

The writer recalls inspecting a house in North London on one occasion following repossession by a building society to find that the borrowers had removed a large number of internal fixtures including all the interior doors, the bath panel and the lavatory seat. On another occasion the occupiers of a small light industrial workshop had stripped out all the electrical wiring and copper plumbing before making their escape.

In law anything permanently affixed to the land becomes part of the security but defaulting mortgagees do not always respect the legal niceties. The mortgage deed may specify that interior and exterior decorations are to be dealt with at specified intervals but few borrowers in financial difficulties will be able to comply.

The mortgagee in possession has an obligation to safeguard the value of the security and obtain a fair price for it when selling bearing in mind that the mortgagor may have a financial interest in the proceeds of sale.

Chapter 18

A typical Building Survey Report

For if the trumpet give an uncertain sound, who shall prepare himself to the battle? So likewise ye, except ye utter by the tongue words easy to be understood, how shall it be known what is spoken?
St Paul

This is a typical report prepared on a house in Chelsea, London SW3, reproduced as an example of a format which might be adopted for the preparation of a report on a simple low rise residential building.

    The report was prepared using a word processor following an inspection which included the taking of written notes on standard field sheets as well as some sketches and also some photography. Photographs are included in the final copy of the report sent to the client and some are reproduced here.

> **SEARCH AND PROBE**
> **Chartered Surveyors**
> **R. E. Search FRICS   I. Probe FRICS**
> **Crumbling Chambers**
> **13 High St.**
> **Weathering**
> **London SW3**
>
> **27th November 1996**

A. Client Esq.
100 Concerned St.
Worrying
Hants.

<u>12 CHELWOOD STREET, CHELSEA, LONDON, SW3</u>

Dear Mr Client,

1. Introduction

I refer to your instruction dated 22nd November 1996, requesting a Building Survey Report on the above property, in accordance with the standard Conditions of Engagement for Building Surveys and Reports which you have seen and signed and a copy of which is annexed to this report.

2.   Conditions of engagement

Standard conditions of engagement for Building Survey Reports were signed by you on 22nd November 1996 and a copy is attached.

2a.   Limitations of inspection

In addition to the limitations set out in the standard Conditions of Engagement no access is currently available to the roof voids and accordingly I have not inspected inside any roof spaces at this stage.

3.   Date of inspection and weather

I inspected the house on 24th November 1996. At the time of my inspection, the property was vacant and unfurnished. The weather was mild and dry.

4.   Conclusion and summary

This house would respond well to a planned programme of re-fitting and redecoration. It is currently neglected and in need of total refurbishment, together with the renewal of services. A number of specific defects were identified during the course of my survey, including dampness, furniture beetle activity and an unsatisfactory timber floor in the basement which needs to be replaced with a different type of floor. If you are in any doubt regarding the likely costs involved in bringing the house up to an acceptable standard of condition, quotations should be obtained before exchange of contracts.

5.   Legal matters

The advice given in this report is prepared on the assumption that the property is Freehold with an unencumbered absolute title and for sale with vacant possession. It is further assumed that the property is not subject to any restrictive covenants which would limit its development for use as a dwelling house, in accordance with your proposals. Further, that all replies to the normal solicitors pre contract searches and enquiries are satisfactory and that the property is not subject to any notices, orders or other matters which could affect its value.

I must emphasise that I have not made any specific enquiries regarding the authorised planning use of the property, which is assumed to be as a single family dwelling house. It should be noted, however, that the accommodation appears to have been sub-divided in the past and may possibly have been in use as separate flats or as a dwelling in multiple occupation.

I recommend that your solicitor makes specific enquiries to confirm whether the property is listed or in a Conservation Area and if so, whether any special rules apply in that regard, or whether the house is listed as being of special architectural or historic interest.

Front elevation viewed from the east

The terrace

The street looking east

6.   General description

Location and amenities

The subject property comprises a four storey, terraced house constructed circa 1880, located in an established residential road, convenient for local amenities.

The accommodation comprises: Hall, 2 Rooms and Kitchen at basement level; Hall, 2 Rooms and Bathroom at ground floor level; Landing, and 3 Rooms at first floor level; and Landing, 2 Rooms and WC and rear roof terrace at second floor level. There is a basement area externally at the front with 3 vaults beneath the pavement. At the rear, there is a small paved, walled garden with shed and external WC. There is no garage or on site car parking. Street parking in this area is by meter and residents permit.

The house faces roughly south, so the sun will generally shine towards the front elevation.

As an appendix to this report I enclose photographs of the house taken at the time of my inspection showing various parts of the structure and certain specific defects discussed in this report.

7.   Structure

7.1   Roofs

There are three roofs. The main roof is of centre valley design and has a modern covering of interlocking concrete tiles on battens and underfelting where seen, draining to a central box gutter which is lined in zincwork. The party walls are raised on each side and there is a rendered parapet wall at the front. The zinc box gutter drains into a uPVC hopper head and rainwater pipe at the rear. I carried out an inspection of this roof, using a ladder, but I was unable to inspect within the two roof voids, due to lack of access. If you buy the house, I recommend that you arrange for two small trap doors to be inserted in the ceilings under the two roof voids to facilitate access, in the future, for maintenance and inspection purposes. In the absence of inspection it is not possible for me to comment on the condition of the roof timbers, the ceilings or to advise on the presence or otherwise of insulation. When access is available, these matters should be checked and if insulation is poor or non existent, I would recommend the provision of 100mm of fibreglass quilt, tucked down neatly between the ceiling joists, to reduce heat losses. As far as I can tell, no ventilation is currently specifically provided for into these roof voids. I would recommend that two tiles on each side be replaced with ventilation tiles and an opening cut through the underfelting (if any) in order to ensure that the roof voids are ventilated. This is necessary in order to ensure that condensation does not occur on the roof timbers, which could lead to rot or other defects.

As viewed externally, the main roof covering was satisfactory when I inspected. There is slight dishing to the tiles but such unevenness is to be expected in a building of this type and age when a fairly heavy covering of concrete tiles is used to replace the original covering, which would have been of slates. Normally, when concrete tiles are used to replace slates, some strengthening of the roof timbers is necessary. I was unable to confirm whether the roof timbers have been strengthened in this case and the point should be checked when access to the roof void is available. It is possible

Zinc valley and interlocking tiles with lead flashings to the main roof-centre valley design

The flat asphalt roof in need of resurfacing and improved balustrades

that some additional strutting to the centre of the roof slopes may be advisable although, at present, the amount of deflection in the form of dishing is minor and not of itself a cause for concern.

The second roof is at a high level over the rear wing and covers the second floor WC off the half landing. This roof has a covering of leadwork, laid to two rolls and draining to a uPVC gutter at the rear. A minor repair is needed to the leadwork at the lip where it is slightly worn, due to foot traffic. Otherwise, this lead roof appears to be in good condition for its age and is not due for immediate renewal. I must emphasise that I have not seen the supporting timbers underneath the leadwork and I am unable to comment regarding their condition. If, in the future, the leadwork is stripped off, then it would obviously be important to have these timbers checked. There is a water tank on top of this lead roof, in a timber housing, encased in green mineral felt. The tank cover was stuck when I inspected and I was unable to lift it to look inside. In general, I would not recommend placing water tanks externally on a roof in this manner, and when the services in this house are renewed I would strongly advise removing the tank and all the external plumbing.

The third roof takes the form of a terrace over the rear first floor room and this appears to be of concrete slab construction. The surface is of asphalt which may have been given additional asphalt coatings, over the years, to extend its life. The asphalt is now extensively cracked and crazed and in my view, now due for re-surfacing. Accordingly, I would recommend that you allow for having all the asphalt surface stripped off and renewed. The asphalt is carried up to form an upstand along the rear wall and to the external door threshold. This detail is poor and has cracked in the past, leaving points where driving rain could enter. At one point, a repair has been made, fairly recently, in mineral felt but this is unlikely to prove durable in the long run and total re-surfacing should include a renewal of the upstands which should be chased into the adjoining brickwork and the detail at the door threshold also needs improvement. Any good reputable asphalting contractor will understand what is required. This terrace has a tubular wrought iron balustrade which is fairly open and would not comply with modern Building Regulations which require that balustrading should be designed so as to not permit the passage of a 100mm diameter sphere. In the interests of safety I would, therefore, recommend that you allow for renewal of the balustrading with wrought ironwork or similar, especially if there may be small children in the house.

In addition to the three roofs there is a shallow balcony at the front of the house over the front porch and bay and providing weather protection for the main entrance. This has decorative wrought iron balustrading and is surfaced in asphalt. The asphalt surface appears fairly modern and is generally satisfactory apart from slight separation at the upstand which needs pointing.

7.2   Chimney stacks, flashing and soakers

Lead flashings are provided to the main roof and at the base of the chimney stacks. Chimney stacks are located on the party walls with two four flue stacks on the left side and one three flue stack on the right side serving this property. Stacks currently have open pots.

If you propose to use chimneys for open fires or appliances, they should first be swept and the condition of the linings checked. In a building of this age, it is likely that the flue would need to be re-lined before it can be safely used. Re-lining would

Front asphalt roof in generally good condition

be essential, if you propose to burn open fires. If flues are to be redundant, I recommend that they be provided with vents at the base and hooded pots at the top, to keep them dry. The front left chimney stack is constructed in fletton brickwork and appears to have been re-built, at some time, in the more recent past. The other two stacks are in stock brickwork or rendering and appear original. Some re-pointing has been carried out in the past and the overall condition of the chimney stacks is reasonable, bearing in mind their age.

7.3   Parapets, parapet gutters, valley gutters

The parapet construction appears satisfactory for its age but ongoing maintenance should be anticipated in the form of periodic re-pointing and repair to rendering, bearing in mind that these walls are very exposed to the weather. The zinc valley gutter in the centre of the main roof is fairly new and in satisfactory condition.

7.4   Gutters, downpipes, gullies where visible

All the rainwater is taken at the rear, mainly through uPVC gutters and rainwater pipes but partly through older cast iron rainwater pipes. Roof water drains to gullies at the back of the house and into a combined foul water and storm water drainage system. The whole system needs an overhaul and some of the uPVC sections are loose and leaking at the joints. Ideally, I would recommend that it all be replaced with modern black uPVC gutters, rainwater pipes and fittings. With any house having solid walls, it is most important to avoid spillage, since this can penetrate directly to the inside.

7.5   Main walls and foundations

I have not undertaken any excavations to examine foundations or subsoil beneath this house. The Geological Survey Map for the area shows subsoils in this locality as being mainly river terrace material, typically alluvial in nature, overlaying London clay. From my own local knowledge I would expect to find the Geological Survey information confirmed but there are considerable local variations in subsoil conditions and the actual conditions beneath this house can only be confirmed by taking a trial bore sample.

Foundations for houses of this age in this locality generally comprise shallow concrete strips with stepped footings, but in the early houses footings were often laid on a bedding of rubble (hoggin) without concrete. Typical foundation depths are often only 300mm to 600mm below basement floor level. Because the house is of basement construction the foundation depth below external ground level will be reasonably good, at least 2.0 metres.

The main walls are constructed in solid brickwork with elevations, mainly in London stock brick but partly in white render at the front and with the lower brickwork painted white at the rear. I carried out a careful examination of the wall surfaces inside and out. I can report finding no signs of unusual movements or settlements to the main walls and the underlying structure appears satisfactory for a building of this type and age. Some re-pointing is required to the external elevations,

Eroded pointing at rear east corner

Eroded pointing at rear west corner

most noticeably at the rear at a high level on each side where the party walls and rear chimney stacks have a brick return elevation. The outer wall to the half landing WC at second floor level rests on two rolled steel joists which span over the room beneath. The outer ends on these rolled steel joists are exposed where they overhang the edge of the asphalt roof – this is poor building practice since steelwork of this type should normally be fully encased as protection against rusting and also to slow the rate of deformation in the event of fire. No adverse effects of the condition of this steelwork were seen when I inspected however. Periodic coating with rust preventative paint to all exposed steel would be advised in the future.

This is an old building constructed in soft lime mortar, so some irregularity to the brick elevations in the form of minor dishing, bulging or slopes to courses must be regarded as normal due to the long-term settlement and shrugging down of the structure over the years. It is not appropriate to judge a building of this age by the standards of a brand new house.

7.6   Damp proof course and sub-floor ventilation

Damp proof courses are hidden behind a rendered plinth at the base of the main walls and cannot be seen without exposure. From my knowledge of houses in this area, I would expect to find that the damp proof courses consist of two layers of slates at the base of the main walls, this being typical for construction of this era. As a general rule, the outside ground level should be 150mm (two brick courses) below the internal floor level or damp proof course, whichever is the lower, with a slope away from the walls to drainage. This is currently not the case and I would recommend that the paving around the rear of the property be lowered to provide the recommended clearance. At present, the rear paving is almost level with the basement kitchen floor. I would also recommend removal of the external WC, the structure of which is bridging the damp proof course.

In the basement the floors are solid in the kitchen, hallway and understairs cupboard. In the two basement living rooms timber floors are provided and I took up floorboards, in order to inspect conditions beneath these timber floors. I can report that the timber floors are badly designed, suffer from furniture beetle activity and have inadequate sub floor ventilation. A number of boards have been replaced in the past, possibly due to rot and this is a problem which could well occur again with a floor of this design. Timbers in the basement timber floors do not appear to be separated from damp oversite and masonry by effective damp proof courses.

If you buy the house I recommend that you allow for replacement of the floors in the two basement living rooms with solid concrete floors. These solid floors should incorporate a damp proof membrane which should be contiguous with the damp proof course in the surrounding walls. All the timbers will need to be removed and the over site area cleared out, so that a solid floor can be laid. I understand that you are proposing to use the front basement room as a kitchen and a solid base would be preferable, in any event, if you wish to have a tiled kitchen floor surface.

7.7   External joinery including window and door frames

This is all generally rather neglected and is now due for complete overhaul and redecoration. Sash cords have recently been replaced in some of the sash windows and the disturbed beadings have not been made good and decorated.

Timber staircase to the front basement area – potentially hazardous and in need of renewal

The front basement area is reached from a gateway and a set of timber steps down from the pavement. The steps are in poor condition with early rot attack to the treads and complete renewal is necessary in the interests of safety. A more durable means of access would be using wrought iron or similar, rather than timber.

7.8   Exterior decorations and paintwork

I recommend that you allow for the preparation and redecoration of all the painted walls, woodwork and ironwork, in due course, in accordance with your own tastes.

8   Internally

8.1   Ceilings, walls and partitions

Most of the ceilings appear to be original lath and plaster construction and are in generally poor condition with extensive unevenness and cracking. In some places, the original ceilings appear to have been replaced. For example, at second floor level the front bedroom has a modern plasterboard ceiling but the rear bedroom has an old lath and plaster ceiling which is noticeably uneven.

I recommend that you allow for renewal of all the original lath and plaster ceilings with modern plasterboard and a plaster skim finish. In those rooms where there are decorative cornices and other features, these will need to be replicated in new plasterwork which will add to the cost. I do not recommend that you retain any of the original lath and plaster.

As regards the internal walls, most of these are timber framed partitions, also clad in lath and plaster and much of this is loose or hollow underneath the existing wallpaper and other linings. Here again, major works are recommended and I would suggest that all the internal stud partitions be re-surfaced in modern plasterboard. Until the old lath and plaster is removed, it is not possible for me to comment on the condition of the timber studs and noggins behind. However, bearing in mind the presence of furniture beetle in timbers in other parts of the house, it seems possible that furniture beetle could be active in the stud partitions, which may need treatment and repair, prior to re-lining with new plasterboard.

At basement and ground floor level, the wall between the front living room and hall appears to be of brickwork. I understand that you wish to remove this wall at basement level. Accordingly, I have examined the construction in this area and I can advise you that the floor joists appear to run front to rear in the living room but not in the hall and the wall in the basement, therefore, supports some floor loading in addition to the wall above. The removal of this wall can be undertaken, provided that a suitable beam is inserted to support the construction above, which should take the form of a rolled steel joist, universal beam or similar, resting on a suitable peer and padstone support at each end and properly encased, in accordance with the requirements of current Building Regulations.

8.2   Fireplaces, flues and chimney breasts

None of the existing fires or fireplaces should be used until the flues have been checked and if necessary relined.

Interior view

Entrance hall

Interior view

8.3   Floors

The timber floor in the two basement living rooms needs replacement, as already mentioned. The solid floors at basement level should provide a suitable base for a new surface covering of tiles or similar. Underneath the stairs the concrete is laid over a polythene damp proof membrane, indicating that this is not an original floor but that it was replaced at some time in the recent past, possibly in response to a problem with rot or beetle attack. It is not possible for me to confirm whether the damp proof membranes are present in all areas, since this can only be ascertained from exposure.

As regards the other timber floors, there is evidence of furniture beetle activity in places but apart from this, in general, they appear satisfactory, bearing in mind the age of the property and the obvious history of past neglect. There is some 'bounce' to the floors in places, which is to be expected in a construction of this type but is not, in my view, excessive.

8.4   Dampness

I tested walls selectively for dampness. Readings were taken on a mild, dry day, following an extended period of dry weather with the interior vacant and unoccupied. In general, the results were satisfactory. However, high readings were obtained to the rear wall in the kitchen, adjoining the external door threshold and in the corner adjacent to the external WC. This appears to be the consequence of bridging of the damp proof course by the high external paving, already mentioned. Some dampness

Floor exposure in basement

Common furniture beetle (Anobium Punctatum) in basement floor

Moisture testing in basement

was also found at the front of the basement hallway, on either side of the external door from the front basement area and here too, local bridging of the damp proof course appears to be the likely cause. On current indications, no major damp proofing works appear to be called for.

8.5   Internal joinery – including doors, staircases and built in fitments

The interior of the house needs complete re-fitting, in accordance with your own tastes and requirements. I do not therefore propose to comment on the existing joinery.

8.6   Internal decorations

The internal decorations need complete renewal, following the re-plastering, already advised.

8.7   Cellars and vaults

There are three vaults under the pavement, used for rough storage, in typical condition. These are not tanked and are unsuitable for any other purpose.

8.8   Infestation rot and other timber defects

I found flight holes of the common furniture beetle in floorboards and sub floor timbers, generally throughout the property and also, to the staircase and understairs cupboard panelling. New flight holes with bore dust indicate that insects have emerged to lay their eggs elsewhere, the results of which can take up to three years to re-appear.

Furniture beetle is active in the basement timber floors, which I have already recommended be replaced. In general, furniture beetle will not thrive in timbers which are completely dry so, if the moisture content in timbers is sufficiently low, the problems may cure itself. If low moisture content cannot be guaranteed, then some treatment of the timbers in this house would be recommended and I suggest, in the first instance, that you obtain a report and quotation from a reputable timber treatment contractor who can issue a long-term guarantee which can be transferred on any subsequent re-sale of the house.

I must emphasise that I have not seen any of the roof timbers either in the flat roofs or in the main roof voids and I am unable to advise you whether furniture beetle or other defects are present in these. Further investigation on this point is, therefore, recommended.

8.9   Thermal insulation

The solid external walls will lose heat more rapidly than modern cavity construction. If there is no ceiling insulation in the roof voids, then heat losses through the roof will be high. The single glazed sliding sash windows are likely to be rather cold and draughty. By modern standards the level of thermal insulation is poor and heating costs will be high.

9   Services

9.1   Electrical

The system needs complete renewal. At present, there are three electric meters and the whole installation is of poor quality. I recommend that a new external meter box be provided in the front basement area and that the whole house be re-wired, in accordance with current Institute of Electrical Engineers' regulations to a modern consumer unit with circuit breakers.

9.2   Gas

The gas service is in poor condition and there are some old gas fires partially disconnected. The system should not be used in its current condition.

I recommend that a new external meter box be provided in the front basement area and that the gas service inside the house be totally replaced.

Views of poor quality interior fittings. These appliances should not be used

9.3   Cold water, plumbing and sanitary fittings

A mixture of lead, copper and iron pipework is provided with an assortment of very dated and poor quality sanitary ware and fittings.

I recommend that you allow for renewal of the incoming mains supply pipe in polythene and that all the plumbing in the house be renewed in modern copper pipework with outlets to new sanitary ware and fittings. The storage tank on the rear flat roof should be removed.

9.4   Hot water and heating

There is currently no central heating system. Hot water is provided from gas water heaters. None of this is satisfactory and I recommend that you allow for the provision of a modern gas fired central heating system, also providing hot water.

9.5   Underground drainage

There are three manholes, two at the rear and one at the front with a private drain running underneath the house and connecting to a sewer in the highway via an interceptor with cleaning eye in the front manhole. The system needs hosing out and some minor repair is required to the manhole benching. The original cleaning eye cover is still in place, on a chain in the front manhole. A visual examination indicates that the drains appear to be in satisfactory condition for their age. However, I have not

Rear manhole

Front manhole with cleaning cover on chain

Slate roof to outbuildings showing slates clipped with tingles

carried out any testing of the pipework under pressure and it was not possible for me to check the drain flow, at the time of my inspection because the water to the house was disconnected. It is important that drains passing beneath buildings are free from leakage and I recommend that a drains test report be obtained prior to exchange of contracts.

10 General

10.1 Garage(s) and outbuilding(s)

The external WC and shed are of poor quality construction with a slate roof. Many of the slates are held in place using tingles. In my view these outbuildings do not add anything to the value of the house and would be better removed.

10.2 The site

There is a small paved, walled garden at the rear with paving in generally poor condition. In my view, the paving around the base of the walls is too high, as already mentioned, and I would strongly recommend that you allow for this to be lowered and that the whole of the courtyard garden be re-landscaped with new paving. The surrounding boundary walls are in fair condition for their type and age and will need routine ongoing maintenance.

10.3 Building regulations, town planning, roads, statutory, mining, environmental matters and services

If you propose to undertake structural alterations to the house in the form of the removal of load bearing walls, consent under Building Regulations will be required and the local District Surveyor should be consulted. Before any works are undertaken, it would also be advisable to confirm whether the house is Listed (I think that this is unlikely) or if it is in a Conservation Area.

11 Valuations

Subject to the results of further tests or investigations recommended in this report proving satisfactory, I form the view that the current open market value of this house in its present condition is £400,000 (four hundred thousand pounds). Suitably refurbished to a standard appropriate to its period and the neighbourhood. I would consider its value to be in the region of £600,000 (six hundred thousand pounds).

Yours faithfully

R. E. Search FRICS

Chapter 19

Legal considerations

The buyer is not entitled to remedy the defect and charge the cost to the surveyor. He is only entitled to damages for breach of contract or for negligence.... So you have to take the difference in valuation.... In other words, how much more did he pay for the house by reason of the negligent report than he would have paid had it been a good report?
Lord Denning MR – Perry v *Sidney Phillips & Son (1982)*

Liability in contract and tort

The chief protection against claims made for negligence in respect of professional work lies in the skill and knowledge which are acquired and maintained by continuous technical study and practical experience. The law requires a standard of care and skill which would be displayed by a reasonably competent person who has the normal skill associated with the profession in question. The law does not require all members of a profession to exhibit the very highest degrees of skill which are to be found among the most eminent practitioners, and it will not normally find a professional man or woman liable for a mere error of judgement on a difficult point.

The standard of care required is, therefore, that standard which a Court would expect to be displayed by a reasonably competent surveyor undertaking surveys and valuations of buildings. If the professional institution issues guidelines for the conduct of this type of work and if the surveyor fails to comply with those guidelines, then this will not, of itself, justify a claim for negligence. However, failure to comply with such guidelines or recommendations may make it more difficult for the surveyor to defend the position should it be necessary to do so in Court.

The standard of care required from all practitioners in a profession is, nevertheless, an exacting one. The law requires that if a surveyor puts himself or herself forward to clients as being qualified and fitted to survey buildings, then he or she should in fact be so qualified and fitted, and capable of achieving reasonable standards. If he or she is unable to carry out particular instructions in cases which involve matters beyond his or her capabilities, then those instructions should be declined.

There is little excuse for a professional man or woman who commits a gross error when advising a client about a particular property. A gross error resulting in substantial and avoidable financial loss to the client is a situation which should never occur. Surveyors will not commit gross errors if they have studied this volume and conducted themselves in accordance with the guidelines and recommendations of their professional institutions, and have carried out careful and thorough inspections and carefully checked their reports before dispatch.

A client may, of course, buy or lease a building, or lend money on it as security and suffer a financial loss in the future due to some unforeseen event, the nature of which no reasonably competent surveyor, carrying out specific instructions of a particular kind, could be expected to foresee. Liability in such circumstances will not arise and the surveyor may have a clear conscience that the loss suffered, while obviously regrettable, was not due to negligence. The only protection for a client against events of an unforeseen nature is insurance in so far as insurance may cover the type of loss involved. It must be emphasised that a surveyor's report is not an insurance policy, and clients should not regard it as such.

It is hoped, therefore, that no reader who is a practising surveyor will ever be in the position of having committed a gross error. There may, however, be a claim from an aggrieved client in respect of some defect in a building or advice given in a report. There may be nothing of substance in the client's claim at all, or the matter may be arguable; the surveyor may find that he has acted for a client who is a 'professional' litigant who may make a habit of engaging in legal disputes of one kind or another. The purpose of this Chapter is to discuss in a general way the present state of the law relating to professional negligence and associated liabilities in England and Wales by reference to the facts of specific cases.

Surveyors should ensure that their liability for professional negligence is covered by a policy placed with a reputable insurer and that the sum insured is adequate. Other insurances should be maintained, depending upon the nature of the surveyor's practice, covering public liability and personal and employee liability. Professional indemnity insurance is compulsory for members of the RICS. The sums insured should be subject to regular review in the light of the changes in property values and building costs generally and the nature of the surveyor's practice in particular. The temptation to under-insure in order to save premium costs should be resisted.

A legal liability which may arise will do so, in English law, under the law of contract where there is either a direct contractual relationship between surveyor and client or a contractual relationship which is held to include a third party; it may also arise under the law of torts where there is no direct contractual relationship but where the law provides that a duty to care exists. The law in both areas has continually evolved since the Second World War due to the establishment of a series of important legal precedents in certain leading cases. The potential liability of a surveyor towards clients and third parties is now much greater than would have been thought possible 40 years ago. In addition, certain limitations have been placed upon the extent to which liability may be avoided by the use of exclusion clauses, particularly by the Unfair Contract Terms Act 1977.

Important and interesting cases

In order to convey the flavour of this process of evolution a number of leading and recent cases are now summarised where the Court judgements may be of interest to surveyors, having particular regard to cases which indicate the Court's reaction to particular circumstances and interpretation of the standard of care required in specific cases. The Unfair Contract Terms Act is also discussed.

Morgan v Perry (1973) 229 EG 1737

This was a case of gross negligence heard on 16 November 1973 before Judge Kenneth Jones in the Queen's Bench Division of the High Court, and it comprised a claim by Mr Morgan against a surveyor Mr Perry in respect of serious faults which developed in a house called 'Samarkand' at Cleeve Hill, Cheltenham.

Mr Morgan had obtained a report from Mr Perry and subsequently purchased the house. The claim was therefore a claim for professional negligence arising out of a contract between the parties, it being alleged that the defendant failed to exercise reasonable care and skill in and about the survey of the house, and the preparation of the report. These allegations were found proved.

The house was built upon a steeply-sloping hillside which, in fact, was so steep that the house was of split-level construction with the living rooms at first floor level at the rear and ground floor level at the front, and the bedrooms at ground floor level at the rear and forming a basement at the front. The house had been built in 1964 and the defendant surveyed the house in July 1968, issuing his report the following day. Under 'Condition external' he reported 'the structure is sound and there is no evidence of settlement'. Under 'Condition internal' he reported 'there are some hairline plaster cracks. These are of a minor nature'. Under the heading 'Opinion' he reported 'We are of the opinion that this property is structurally sound. There should be no undue maintenance costs in the foreseeable future'.

The Judge commented that the report was a remarkable document; apart from the minor hairline cracking to plaster in unspecified locations, no defect of any kind was referred to. The defendant had given the house a completely clean bill of health and the plaintiff had proceeded with the purchase. However, while laying the drawing-room carpet the plaintiff had noted that the floor was not level. Moreover, after taking occupation he had noted certain cracks in the external walls. Lacking any expert knowledge the plaintiff, at that time, had not appreciated the significance of the sloping floor or the cracks.

In May 1969 the plaintiff applied for a mortgage, and in June 1969 a survey had been undertaken on behalf of the Halifax Building Society. The Halifax surveyor found evidence of settlement, and eventually, having failed to obtain subsidence insurance, the Halifax declined to advance any money on the security of the house. In due course bore holes were sunk and advice taken. A firm of consulting engineers reported in January 1972.

Engineers reported that the topsoil to a depth of 3.3 metres was slipping on the consolidated clay beneath. The slip was taking place in the three masses; one mass was slipping rotationally; one mass was slipping so that the garage was tilting towards the house; and a general movement down the hillside was taking place. The engineers advised demolition of the house, slope stabilisation work, special foundations and a new house, this being the only way to produce a house on the site with firm and lasting foundations.

The Court decided that it was fundamental to establish the visible condition of the house in July 1968 when the original survey took place. Twelve witnesses were called to give evidence, some by the plaintiff and some by the defendant. The original owner, and vendor to the plaintiff, was called to give evidence and the Judge found that he was less than frank with the Court regarding defects which apparently arose during his period of ownership, and some filling of cracks which had been undertaken prior to sale. Whether or not the vendor had deliberately covered up the defects, the Judge found that the effect of crack filling, replastering and wallpapering which had been undertaken was the same as that of a deliberate cover-up job with the intention of making the house saleable. The Judge decided that the defendant, Perry, did not in fact observe any defects when he carried out his survey, but he found it impossible to rely upon his evidence that no defects were observable.

The Judge found that the defendant had failed to notice repointing, or the shape and extent of various cracks, repairs to plasterwork, doors not closing properly, over-painting and so forth. He commented that, human nature being what it was, there would be instances of vendors taking steps to cover up defects, and that was at least part of the reason why a prospective purchaser employed a surveyor. Considered against the fact that the house was built on a steeply-sloping site with known potential dangers, the surveyor should have undertaken a close examination of the house, which would have revealed the slopes to the floors. Evidence was given that the highway outside the house had been subsiding for many years, and the surveyor should have looked at the highway, which would have aroused his suspicions and led him to make enquiries with the highway authority. The Judge found that the defendant committed a breach of his contract with the plaintiff and provided him with a report which was so misleading as to be valueless. Accordingly, an award of substantial damages plus costs was made.

Leigh v Unsworth (1972) 230 EG 501

A case rather similar to *Morgan* v *Perry*, but with a different outcome, was heard before Judge Everett QC in the Queen's Bench Division of the High Court on 21 November 1972. This comprised a claim by Mr Leigh against a surveyor Mr Unsworth in respect of faults in a house called 'Greenacres' at Swallowfield, Berkshire.

The defendant surveyed the house in April 1965 and the plaintiff subsequently purchased. The writ aginst the defendant was not issued until 5 years and 51 weeks after the date of the survey, a matter which the Judge described as regrettable, especially as a further delay then ensued before the matter was finally brought to trial, so that 7½ years elapsed between the date of the survey and the date of the trial.

The house had apparently been built in about 1960 to a very poor standard, especially in respect of the main defect which was that the outer walls rested on conventional foundations but the internal walls rested on a concrete raft supported by a backfilling of loose clay and other unsuitable material, rather than conventional hardcore. It was agreed at the trial that the manner in which support was provided for the internal walls was lamentable building practice. The issue was whether, at the time of the survey, there were present sufficient indications of the existence of settlement of the internal walls and raft, such as to make a competent surveyor advise his client of the problem.

The Judge held that the duty of the surveyor in these circumstances falls into two areas. The first is actual observation, and the inferences to be drawn from the results of that observation. The second is just what is required of him by way of information, put quite broadly, to his client. In the report the defendant had stated 'Generally the property is soundly constructed and no major defects were found.

The works of maintenance and repair are generally maintenance works due to original shrinkage and drying out'. He also stated that the chimney stacks had lead flashings, whereas in fact they had cement fillets which subsequently leaked. The report stated that 'The roof was found to be strongly framed with good-quality timbers of ample-sized scantlings' and 'the ceilings were covered with fibreglass except for a proportion of the roof which was not so treated'. The defendant claimed that the main roof was poorly constructed, in that there was a complete lack of vertical hangers, and purlins were not properly strutted having individual lengths butt-jointed and cantilevered. Further that, apart from one small area, the ceilings were not in fact insulated. The defendant had rendered an account for builder's charges in connection with drains tests, but, it was alleged, had failed to discover a massive leak in a drain under the building.

Having moved into the building the plaintiff discovered various defects, including the fact that the kitchen floor had sunk between 12 and 18 mm. During the course of the trial evidence was given from each side as to the condition of the premises as it had developed, which was agreed to amount to a serious settlement, and the condition of the premises as would have existed at the time of the survey. The outcome was that the Judge found that three minor errors in the report had been made: the poor construction of the roof timbers; the lack of lead flashing to the stack; and the poor level of insulation.

As to the substance of the case, which was the settlement of the internal raft and walls, the Judge found that the plaintiff's case had not been made out, and an award for the three minor matters only, plus some costs, was made. The leakage of the drains was considered to have arisen after the date of the survey, and the defendant's evidence, that the drains passed a test at the time of the survey, was accepted.

Clearly, if the three minor errors in the report had not been made the outcome of the case would have been to vindicate the surveyor, in the circumstances described. The significance of the fact that the writ was served 5 years and 51 weeks after the date of the survey lies in the fact that an action for breach of contract must be commenced within six years of the date of the alleged breach. In order to be able to defend an action for breach of contract all records relating to the matter must be retained for at least six years. In respect of actions in tort the time limit may, in effect, be longer since such actions may be commenced within six years of damage resulting to the building (see *Pirelli General Cable Works Ltd.* v *Oscar Faber & Partners [1983] 1 All ER 65, [1983] 2 WLR 6*). For this reason it is suggested that all surveyors' records dealing with surveys and reports be retained, in their entirety, without time limit.

Hedley Byrne & Co. Ltd. v Heller & Partners Ltd. [1963] 2 All ER 575, [1964] AC 465

In Hedley Byrne the House of Lords introduced a new principle into the common law which was that a liability arose in tort between the maker of a statement of fact and opinion and a person who may rely upon that statement of fact or opinion, even if there is no direct contractual relationship between them. Prior to Hedley Byrne the Courts had restricted liability in tort to actions, rather than the spoken or written word, and restricted the right to compensation for damage to that occasioned by physical damage rather than financial loss.

Hedley Byrne established a new type of liability in tort as being that for the spoken or written word giving rise to financial loss where the writer or speaker and the person who relies upon the advice have a relationship of proximity or neighbourhood which gives rise to a duty to care.

Dutton v Bognor Regis Urban District Council [1972] 1 All ER 462, [1972] 1 QB 373

In *Dutton* v *Bognor* the question of liability in tort was further considered, and extended. In this case the plaintiff, Mrs Dutton, purchased a house in 1960 not knowing that it had been built on made ground, in fact an old rubbish tip. It became clear, shortly after the purchase, that the house was unstable, and the plaintiff sued the local authority for negligence and breach of statutory duty in respect of the inspection of the house when under construction pursuant to their powers under the Public Health Acts and local bye-laws (since superseded by the Building Regulations). The trial Judge found that the local authority surveyor had been negligent in not noticing evidence of rubbish in the foundation trenches before they were covered-up.

Murphy v Brentwood UDC and DOE v Thomas Bates and Son [House of Lords – 26 July 1990]

In these two important later cases the position established in *Dutton* v *Bognor* [supra] and confirmed in *Anns* v *Merton LBC* [1978] was overruled. These cases were found to have been wrongly decided as regards their statutory functions. The result of these most recent cases appears to be that local authority liability extends to physical injury to persons or property on account of defective works but not the defective buildings themselves.

Yianni v Edwin Evans & Sons (a firm) [1981] 3 All ER 592, [1981] 3 WLR 843

In the High Court Mr Justice Park ruled that there was a sufficient relationship of proximity between the surveyor to a building society and the borrowers that the borrowers could sue the surveyors in tort, notwithstanding disclaimers of liability used by the building society.

In 1975 Mr and Mrs Yianni, as sitting tenants, were offered a house in North London at a price of £15 000. They applied to the Halifax Building Society for a mortgage of £12 000 which was an advance of 80% needed in order that they could buy. The Halifax instructed their surveyors, Edwin Evans & Sons, to prepare a pro-forma report and valuation of the house in the normal way. In their offer of a mortgage advance the Halifax enclosed an explanatory booklet which included a statement that the Building Society accepted no liability for the accuracy of the valuation, and that the valuer's report was confidential information used by the Society to determine whether or not an advance should be made. It was stated that if the buyers required a survey they should instruct an independent surveyor, and indeed this was recommended. The Yiannis did not read the booklet, and the report by Edwin Evans & Sons did not reveal any faults. In due course the Halifax lent £12 000 as an 80% advance indicating that their valuation was £15 000.

It transpired that some cracks had apparently appeared in the house before it had been offered to the plaintiffs, and the then owners had discovered subsidence. Some repairs and redecorations had been carried out prior to the survey. After the purchase further cracks appeared and were diagnosed as resulting from subsidence; it was found that the end wall of the house needed rebuilding, and other walls needed underpinning. By 1978 the estimated cost of repairs had risen to £18 000. Edwin Evans admitted negligence in failing to notice that the building had been subject to past subsidence, in failing to take proper steps to have. reasons for subsidence investigated, and in reporting to the Halifax that it was suitable security for an 80% loan. They claimed, however, that they did not owe any duty to care to the plaintiffs, with whom they had no contractual relationship and with whom there was no liability in tort.

For their part the plaintiffs contended that Edwin Evans owed a duty of care to them since they relied upon the Halifax offer as indicating that the property was worth £15 000. The Judge agreed that a relationship of proximity existed such that, in the reasonable contemplation of the defendants, carelessness on their part might be likely to cause damage to the plaintiffs.

During the course of the hearing evidence was given that perhaps only 10% of house purchasers obtain an independent surveyor's report of their own, the remainder relying on the building society surveyor or their own inspections as being sufficient indication that the property is sound and worth the price. In these circumstances the Judge held that the plaintiffs were not liable for contributory negligence in failing to obtain their own report, and the defendants were found liable for substantial damages. On the facts of this case at least the Courts have decided that a purchaser is right to place some reliance upon the building society report as an indication of the condition, and value, of the property. The implications will be important for the future conduct of building society inspections (see Chapter 11).

Donoghue (or McAlister) *v* Stevenson [1932] All ER Rep 1, [1932] AC 562

It will be seen from the summaries of the foregoing Hedley Byrne, Dutton and Yianni cases that the range of persons to whom a surveyor may be held responsible in tort is wide, and has progressively widened in recent years. The process by which the Courts can make new law in this way is termed the doctrine of *stare decisis*, by which points of law are determined by the Courts so as to establish precedents, the decisions of the Appeal Courts having greater weight than those of the High Court, with the decisions of the House of Lords, the premier appelate court (the European Courts excepted), having the highest stand.

In *Donoghue* v *Stevenson* Lord Atkin described the range of people to whom a duty to care is owed, under the law of torts, in the following terms: 'The rule that you are to love your neighbour becomes, in law, you must not injure your neighbour and the lawyers' question "who is my neighbour"? receives a restricted reply. You must take reasonable care to avoid acts or omissions which you can reasonably foresee would be likely to injure your neighbour. Who then, in law, is my neighbour? The answer seems to be – persons who are so closely and directly affected by my act that I ought reasonably to have them in contemplation as being affected when I am directing my mind to the acts or omissions which are called in question'.

Bolam *v* Friern Hospital Management Committee [1957] 2 All ER 118, [1957] 1 WLR 582

In this case the standard of care required from a professional man was described by McNair J. as 'The test is that of the ordinary skilled man exercising and professing to have that special skill. A man need not possess the highest expert skill: it is well established law that it is sufficient if he exercises skill of an ordinary competent man exercising that particular art'.

Rona *v* Pearce (1953) 162 EG 380

In this case the extent to which a surveyor should describe and comment upon aspects of a building and the action he should take in respect of those matters which could not be inspected or confirmed was considered by Hilbery J., who said 'It is highly important to ordinary members of the lay public that a surveyor should use proper care to warn them regarding matters about which they should be warned on the construction or otherwise of a piece of property and that they should be told what are the facts. If surveyor cannot ascertain any fact he should say he was unable to ascertain it and therefore unable to pass any opinion on it'.

Perry *v* Sidney Phillips & Son (a firm) [1982] 1 All ER 1005, [1982] 3 All ER 705, [1982] 1 WLR 1297

This case, which was a claim arising out of a negligent survey, was subject to hearings in the High Court and subsequently in the Court of Appeal. The case is of interest because of the facts, and because the issue of the quantum of damages was considered; the issue on damages being whether these should be assessed on the basis of diminution of value, or cost of necessary repairs, and the date at which damages should be assessed.

In 1976 Mr Perry bought a cottage in Worcestershire for £27 000, having first obtained a report from surveyors Sidney Phillips & Son in which the latter reported that the cottage was generally in order and a reasonable buy at the price. Subsequently Mr Perry discovered a number of defects and commenced an action for breach of contract and negligence.

In the High Court the main allegations were either conceded by the defendants, or upheld by the Judge. On the understanding that Mr Perry intended to remain in the house for the foreseeable future the High Court Judge awarded damages for (i) the cost of repairing defects which the surveyors had negligently overlooked; (ii) compensation for the inconvenience and discomfort anticipated while the remedial works were being carried out, and (iii) compensation for distress, discomfort and other consequences of living in a house in defective condition. The High Court Judge further held that repair costs were to be assessed at the date of judgement, rather than the date of the breach of contract since the defendant was unable to undertake the works at an earlier date due to lack of funds. The High Court judgement on negligence was accepted by the defendants, but they appealed to the Court of Appeal on the issue of the assessment of damages.

The facts, which were not at issue in the Appeal Court, were set out in the High Court judgement ([1982] 1 All ER 1005). The defects in the property, in respect of which a lack of care in the undertaking of the survey and preparation of the report was established, included a number of defects in the roof, a blocked doorway, lack of weatherproofing to windows, a chimney stack lacking proper support, a bowed and bulging wall, water penetration to the interior and various minor defects including a loose basin and loose wall tiles. In addition, the cottage had septic tank drainage which was unsatisfactory and not functioning correctly; the drainage system was a nuisance and generated unpleasant smells.

Unfortunately, by the time this case reached the Court of Appeal the plaintiff's circumstances had changed and he had sold the cottage; it therefore followed that it was no longer possible for the plaintiff to argue the merits of 'cost of repair' as an alternative to 'diminution in value'. The original award of damages based upon cost of repair was made only on the understanding that the plaintiff intended to reside in the house as his home for the foreseeable future.

The Appeal Court reduced the damages to a figure equal to the difference between the price paid for the house, and its true value, as at the date of the purchase, plus interest as some compensation for inflation and 'vexation'; compensation for vexation being a modest recognition of the anxiety, worry and distress arising from the physical consequences of the breach. Damages were to be assessed by the official referee. The surveyors' appeal was therefore allowed on the principle that damages are assessed on the basis of diminution in value, and as at the date of the breach. Two members of the Appeal Court were careful to emphasise that the changed circumstances prompted the decision, and that 'cost of repair' was, at the time of the Appeal Court hearing, no longer arguable. The third member, Lord Denning MR, was however prepared to venture a more general opinion of the position, and he said: 'Damages are to be assessed at the time of the breach, according to the difference in price which the buyer would have given if the report had been carefully made from that which he in fact gave owing to the negligence of the surveyor.... The buyer is not entitled to remedy the defect and charge the cost to the surveyor. He is only entitled to damages for breach of contract or for negligence.... So you have to take the difference in valuation.... In other words, how much more did he pay for the house by reason of the negligent report than he would have paid had it been a good report?'.

Lord Denning added that, while damages would be assessed at the date of the breach, they would carry interest, so that by this means some protection against inflation was provided.

The Appeal Court was not required to assess damages. However, the principle of damages for vexation was confirmed, and some indication given that such sums as may be awarded under this heading should not be large; Lord Denning referred to 'modest compensation', Oliver LJ suggested that such damages would be 'not very substantial', and Kerr LJ confirmed that vexation damages were limited and not intended to deal with 'tension or frustration of a person involved in a legal dispute.... such aggravation is experienced by almost all litigants'. Vexation damages were intended to deal only with the physical consequences of the surveyors' breach of contract.

Eames London Estates Ltd. v North Hertfordshire District Council (1981) 259 EG 491

In the event that defects arise in the construction of a building, a number of parties may be at fault, apart from the surveyor who may have inspected the building for the purchaser. Rules governing the extent of liability have been stated in a number of post-War legal actions, in addition to those already mentioned. On occasion, local authority building control, architects, consulting engineers, builders, developers and surveyors have all been found liable of breach of duty to care, breach of statutory duty, or negligence. For a broad view the reader is recommended to study the judgements in *Anns* v *Merton London Borough Council* [1977] 2 All ER 492, [1978] AC 728: *Batty* v *Metropolitan Property Realisations* [1978] 2 All ER 445, [1978] QB 554; *Acrescrest Ltd.* v *W. S. Hattrell & Partners* (a firm) [1983] 1 All ER 17, [1982] 3 WLR 1076 and *Eames London Estates Ltd.* v *North Hertforshire District Council* (1981) 259 EG 491.

In Eames the case concerned an industrial building which was designed by an architect, built by a firm of contractors for a firm of developers, and subsequently let with the freehold reversionary interest sold to investors. The building was erected partly on made ground subject to past landfilling, and some cracks appeared in 1971, but it was not until 1976 that the cracking caused such alarm that expert advice was sought by the tenants.

Eventually a total of three plaintiffs took action against four defendants, the defendants being the architect, the local authority who passed the foundations, the developers and the builders. Judge Fay held that all four defendants were liable to pay damages. The architect was considered to bear a high degree of responsibility, since the building, and its foundations, were his design; the fact that the local authority had passed the design was no excuse. The architect was found liable to pay 32½% of the total damages awarded. The developers were held liable for 22½% of the damages having regard to their apparent knowledge of the site conditions, and lack of suitable instruction as to foundation design. The builders were held 22½% liable, despite evidence that they had found fill material in the trenches and reported this to the architect, and proceeded only on the architect's instructions. The local authority were also held 22½% liable for failing to take sufficient steps to confirm that the foundation design was suitable.

Eames is of interest because of the manner in which liability was apportioned among the negligent parties. The developers could not evade liability by pointing to the fact that they had consulted an architect; they were still liable to ensure that the land was suitable for the foundations proposed (following *Batty* v *Metropolitan Property Realisations* (supra)). The builders were unable to plead successfully that they were acting on the instructions of the architect and building inspector from the local authority in continuing to lay foundations in bad ground.

Eames also touched upon the question of the time from which the period of six years prescribed in the limitation Act for actions in tort or contract would run in circumstances where progressive crack damage was appearing in the building. It was held that this period ran, for the purpose of an action in tort, from that time when reasonable skill and diligence would have enabled the plaintiffs to have detected not merely the symptoms (cracks) but also the nature of the underlying cause (differential settlement). In Eames the plaintiffs' case against the defendants was held to be within the necessary time limit, having regard to the evidence given as to the nature of the structural movements involved and the dates at which crack damage became apparent. Interestingly, the architect claimed to have known in 1972 that differential settlement was probably the cause, this being some time before the plaintiffs were held to have appreciated this fact; it was held, however, that the architect could not rely upon his own knowledge to defeat a claim made against him on the grounds that the claim was therefore out-of-time.

Balcomb *v* Wards Construction (Medway) Ltd.; Pethybridge *v* Wards Construction (Medway) Ltd. (1981) 259 EG 765

In the Balcombe and Pethybridge cases two claims were heard together, both being claims by the purchasers of houses against the builders and their consulting engineers arising out of cracking in newly-built houses resulting from heave damage, the buildings having been erected with shallow foundations on a London clay site where a number of trees had been removed prior to commencement of construction.

Evidence was given that, following the removal of trees, heave damage can last for many years, perhaps as many as ten years, and that the shallow foundations used at a reputed depth of 1.05 metres were unsuitable. Engineers had advised the use of foundations at this depth, suggesting that it would be deep enough to avoid damage from any change in the condition of the clay. The houses had been built in 1972, and in 1974 substantial repairs were undertaken; this did not cure the problem and it was agreed, for the purpose of the legal action, that the cost of proper repair exceeded the value of the houses.

The builders, Wards, admitted liability in contract to the purchasers and agreed the quantum of damages during the course of the trial. The consulting engineers also agreed the quantum of damages if they were liable to the purchasers or the builders, but they disputed the issue of liability. The Judge, Sir Douglas Frank, held that the engineers were liable to the builders for breach of contract in failing to exercise professional skill, and also liable to the builders in tort for breach of duty to care. The Judge also held the engineers liable directly to the purchasers in tort following *Bolam* v *Friern Hospital Management Committee* (supra) and *Batty* v *Metropolitan Property Realisations* (supra).

The Judge considered, on the evidence, that in 1971 when the engineers took soil samples and gave their advice to the builders a competent engineer, encountering London clay, would have made enquiries to discover whether or not there had been trees on the site and, finding that there had been, would have caused moisture content and plasticity tests to be made on the clay. It was considered that the results of such tests would have necessitated different advice from that actually given in that deeper and special foundations would have been required to accommodate future clay swell.

Interestingly it was held that the builders were not themselves liable for negligence; the Judge considered that the builders discharged their duty by employing engineers to advise them, in this respect differing from the Eames case (supra) where the employment of an architect was not considered sufficient to absolve either the developer or the builder from a measure of responsibility for defective construction. Having regard to the fact that Wards were not themselves qualified to carry out the site investigations the Judge found it impossible to think of any other way that they could have discharged their duty other than to rely upon the consulting engineers' advice; accordingly no contributory negligence by the builders was found, and the full burden of liability for the defective foundations rested with the engineers.

Surveyors are advised to foster a relationship with a firm of consulting engineers, and to refer to them for advice on structural matters in many situations (see Chapter 3). It is suggested that, in view of the possibility that a shared liability may arise for advice given, all requests for assistance from consulting engineers should be made in writing stating the specific services required and requesting that engineers themselves should advise on the nature and extent of any testing or sampling they require to provide complete advice and that any limitations involved be carefully explained, in writing, to the client.

McGuirk *v* Homes (Basildon) Ltd. and French Kier Holdings Ltd. (*Estates Times* 29 January 1982)

In this case a claim was made by the purchasers of a house built with mortar found to consist of only one part Portland cement to thirteen parts of sand. The very weak mortar was stated by a structural engineer, called by the defendants, to have complied with building bye-laws applicable at the time, and with BSI CP 111:1970, and the defendants denied any breach of duty.

The Judge found differently, in that the duty of the defendants was 'to build in a substantial and workmanlike manner, and not merely to comply with the bye-laws'. The house had been built in 1960 and the local bye-laws applicable at the time were those of Basildon Urban District Council dated 1958. After 1965 the bye-laws were replaced by the Building Regulations in England Wales (outside inner London), and BS Code of Practice CP 111 has been revised, so that it is thought unlikely that 1:13 mortar could be defended today on the grounds that it complied.

The defect was not discovered for some years. In 1978 it was noticed that smoke was escaping from the chimney stack through the mortar joints, and a local builder was asked to investigate. He found the mortar between the bricks being blown away by the wind and many bricks loose so that they could be lifted out. Further investigation by the plaintiff revealed that the problem was quite general and applied to all the brickwork in the house.

The house has been purchased in 1960 for £3165. The eventual award of damages for diminution in value and loss of amenity was £6684.92. Concluding his judgement Judge Leonard said that the case had been a misfortune for both sides, in that the defendants found themselves saddled with an unexpected liability over 20 years after they had sold the house, and he added 'I regard this as a misfortune for them because I think it is most improbable that anyone in authority intended an unduly weak mortar mix to be used, or suspected that it was being used. The saving in expense would be minimal. It is no part of my function to speculate on the explanation. It is enough for one to say that nothing in the case has come to light suggesting any deliberate intention on the part of the defendants to build otherwise than in accordance with the contract. No word has been said suggesting that the rest of the house was not properly and soundly built'.

It is worth emphasising that a surveyor inspecting a building under construction can undertake only spot checks on materials and workmanship. The modern practice of dispensing with an on-site clerk of works (who represents the client or purchaser rather than the contractor), and the impracticability of undertaking continuous on-site supervision, means that, no matter how frequent and diligent are the surveyor's inspections, they cannot prevent deliberate breaches of good building practice by the workforce occurring between site visits. If workmen on site wish there are many opportunities for them to undertake work badly, cover up the evidence and avoid detection. Some indication of the overall quality of the workmanship and the skill and honesty of the tradesmen on site can, however, be gleaned from meeting and talking with the workforce (see Chapter 14).

London & South of England Building Society *v* Stone (1981) 261 EG 463

Mr Stone prepared a pro-forma building society report on a house in Wiltshire and valued it at £14 850, the society then lent £11 880 to the purchasers Mr and Mrs

Robinson. The house subsided, and remedial works eventually cost £29 000. The society decided to waive their rights under the mortgage deed against Mr and Mrs Robinson and proceeded against Mr Stone for £29 000 plus costs. Mr Justice Russell did not award the cost of remedial works as damages, but adopted a lower figure of £11 880 as being the difference between the stated value, and actual value of the property at the time of the report, and further deducted £3000 from this because he considered that the society had failed to mitigate their loss by waiving their rights against the borrowers. A net award of £8880 plus some costs was made.

The judgement recorded that the surveyor, Mr Stone, was a frank and honest witness who displayed a professional approach to the allegations made and emerged from the witness box with credit. He described his inspection, lasting about an hour, during which he had not seen anything untoward.

The case against Mr Stone was that he should have noticed an important crack, even allowing for some cosmetic remedial works which had been undertaken, and that he should have detected rotational movement in an outhouse, and that he should have noted sticking doors and cracking inside the house. Prior to his survey the then owner had apparently made an insurance claim for subsidence damage, which had been resisted by the insurers on the ground that it originated before they came on risk. The house had been built in about 1961 on sloping ground and on the site of an old quarry. Records existed in the form of a surveyor's report and photographs prepared for the insurers, indicating the condition in February 1976. The inspection by Mr Stone was in May 1976, so that considerable evidence existed of a major structural fault as at the date of Mr Stone's inspection.

The reader will not need reminding by now that I have emphasised in this book the desirability of acquiring local knowledge about old quarry workings and the like, and the higher risks of structural damage likely to arise with buildings erected on sloping, as opposed to level, sites which should place a surveyor on guard.

The Judge noted that attempts had been made by the vendor to better his property and that this would have resulted in some covering up of crack damage. It was felt that evidence of past internal damage was not available to be seen when Mr Stone inspected. The Judge was, however, driven to the conclusion that the crack in the exterior of the structure, and the rotational movement in the outhouse, were available to be seen and that, for whatever reason, Mr Stone had failed to detect them. The Judge found this to be an isolated lapse on the part of a surveyor of considerable expertise and experience, but that liability was plainly established.

The Judge commented upon the nature and extent of a building society survey; that it was not a full structural survey but merely a survey for building society valuation purposes. Nevertheless, the surveyor in this instance had failed to exercise the degree of care and skill expected of a reasonably competent surveyor carrying out the type of examination that he was required to carry out on behalf of the building society.

Fryer v Bunney (1982) 263 EG 158

An action for negligence and breach of duty to care was brought against a surveyor arising out of serious dampness found in a house due to leakage within solid floors from defective small-bore central heating pipework. The defendant surveyor, Mr Bunney, reported that the house had been checked with a moisture meter and that no reading of dampness had been found. When the purchasers took possession they soon discovered serious dampness, which was confirmed by measuring the loss of water from the central heating header tank. It became necessary to excavate the hall floor, remove defective pipework and renew joints. The judgement went in favour of the plaintiffs and the defendant surveyor was held to have been negligent in not noticing or recording the dampness. Damages of £5346 were assessed, comprising cost of repairs and decorations, reduced value of shrunken carpets, and an amount for vexation arising out of the plaintiff's inconvenience and distress (the cost of remedying the defects being presumably equivalent to the diminution in value of the property).

Problems associated with small-bore and micro-bore pipework embedded in solid floors are well recognised. It is essential that such pipework be tested for leaks under pressure before screeds are applied; frequently this is not done. In *Fryer* v *Bunney* the evidence was that soldered joints had not been properly made, and pipes were merely a push fit (indicating, among other things, the contractor's negligence). It is also essential that lagging be properly provided to permit thermal expansion and contraction. Even well-made joints can fail if the pipework is screeded in solidly and unable to respond to thermal movements. From a surveyor's point of view such pipework is inaccessible on subsequent survey, and the only practicable tests are of moisture levels in surrounding walls and floors. Unfortunately for Mr Bunney it was held that he had not undertaken such testing to the standard and extent required in the circumstances.

Fisher *v* Knowles (1982) 262 EG 1083

The purchasers claimed against their surveyor some four years after the purchase, particularising 14 items as being alleged defects; the judge found that the surveyor had failed to warn his clients of only two matters, being some wet-rot to window frames and defects to ceiling joists.

Instructions arose because the surveyor was to inspect the house for the building society, and the purchasers approached him with a request that he also provide them with a private report and valuation. The Judge found that these instructions amounted to something more than a bare valuation report, but something less than a full building survey. The RICS Practice Nopte (which was not published when the allegedly negligent inspection was made) now recommends that if something less than a full building survey is to be carried out it is important that the true nature and extent of the obligation being assumed by the surveyor is understood by, and agreed with, the client.

The Judge emphasised that the burden is on the plaintiffs to prove that, on the balance of the probabilities, the defects were of a kind to which the defendant should have drawn attention. Making the best he could of the evidence the Judge awarded £500 damages for diminution in value of the house, resulting from the two alleged defects found to have existed at the time of the inspection and missed by the surveyor. A claim for a further 12 items was unsuccessful.

Wet-rot to softwood window frames is a known hazard, and the defendant surveyor was in the habit of testing joinery with a penknife; he considered that he would have done so in this case. The Judge found, on balance, that there had been wet-rot to the window frames, and that for some reason the surveyor had failed to detect it. He also felt that the surveyor had failed to notice deflecting ceilings under a bathroom floor due to the poor fixing of a WC pan, which would have placed him on guard.

The writer would recommend that particular care be taken in inspecting exterior painted softwood for signs of rot, and also whenever possible in checking conditions under WC pans resting on timber floors, since rot to timbers from leaking seals will commonly be found.

Treml v Ernest W. Gibson & Partners (1984) 272 Estates Gazette 68

Ernest Gibson, a member of the Incorporated Society of Valuers and Auctioneers, was held liable for substantial damages in this case which illustrates how damages can mount up if serious building defects are found. The facts, of considerable interest to practitioners, were briefly as follows.

Mrs Treml had agreed to buy No 9 Keens Road, Croydon, Surrey in June 1979 for £21,000, subject to survey. Mr Gibson was instructed and reported on this small terraced house finding no obvious defects but advising some treatment of woodworm and rewiring; Mrs Treml bought the house and moved in.

Unfortunately Mr Gibson had failed to notice that the house lacked fire walls in the roof void and had a modern concrete tile covering which was much heavier than the original slate covering. The roof trusses were thrusting against the front and rear walls. In time bulges appeared in those walls as the thrust increased until shoring and structural support was needed to prevent collapse – in October 1982 the local authority actually issued a Dangerous Structure notice.

Mrs Treml asked Mr Gibson to return to the house in 1980, 1981 and 1982 because of various problems but Mr Gibson failed to notice the fundamental defect. Eventually Mrs Treml took expert advice and the nature of the problem became apparent. By the time the case came to court the defendant conceded negligence and the only question to be determined became that of damages.

Mr Justice Popplewell said that following the decision in Perry v Sidney Phillips & Son (supra) the proper method to assess damages was to calculate the difference in value between what the plaintiff actually paid and what the real value of the house was, and to add interest plus any special damages for vexation and inconvenience, and the temporary costs involved which arose in this case because the plaintiff would have to move out of the house for a time whilst it was repaired. Evidence that the real value was £8,000 was accepted, leaving a sum of £13,000 as damages for the negligent survey plus interest and vexation damages.

Mrs Treml had also claimed for the cost of employing engineers, surveyors and builders to arrange emergency support and advise on the nature of the defect. The builders charged £756, the engineers £1,255 and the surveyors £115. The judge did not allow engineers costs apart from £50 relating to the report dealing with the question of the defendant's negligence, the reason being that having assessed damages in relation to the market value of the property it would be unfair also to add further damages which were in the nature of repair costs as this would result in the plaintiff securing compensation in excess of the actual damage suffered.

As an additional complication it transpired that grant assistance was available under the 1974 Housing Act from the local authority in respect of a proportion of repair costs. The Judge held that the possibility of grant aid being available was irrelevant to a claim where the measure of damage was the difference in value, a point strongly argued by counsel for both sides.

Further damages were awarded for a number of incidental costs including £3,850 for the cost of 70 days at a hotel at £55 per day for Mrs Treml and her family whilst the work was done, plus £1,000 for meals and £50 for laundry. A further £1,248 was awarded for the costs of furniture removal and storage, and accommodation for Mrs Treml's cat. £1,250 for vexation and inconvenience had been agreed by the parties. Total damages came to £32,750 plus interest of £8,500. The judge ordered a stay on £6,000 of these damages to cover the possibility that the defendant might appeal, on a point of law, that the local authority grant should have been deducted.

Damages at the end of the day were much higher than they might have been had remedial work been carried out at an early stage, but the court recognised that the plaintiff could not in fact carry out any work until she had successfully sued the defendant, because she did not have the money. Moreover, raising finance would have been difficult on her income and with no assets save this defective house. This being a terraced house there was also some disagreement initially between neighbours on the nature of the works required.

It is suggested that a surveyor should always inspect the roof void if this is accessible. A lack of party walls carried up to provide fire walls in terraced roof voids will commonly be encountered in terraces dating from the last century and this must always be commented upon. If light coverings, such as slates, are replaced with heavier coverings, such as concrete tiles, it is essential that a structural check be first made to ensure that the timbers will take the higher loads – this is often not done resulting in roof sag or roof spread. Consent under Building Regulations or London Building By-Laws is generally required for work involving a heavier roof covering – but roofing contractors do not apply in many cases.

Stevenson v Nationwide Building Society (1984) 25 July 1984

A staff valuer employed by the building society inspected a property which consisted of two shops, a maisonette and a flat built over a river and supported on steel girders and an infilling of concrete. The plaintiff was an estate agent who was purchasing the premises with the assistance of a loan from the society.

When the plaintiff made the application for the loan he completed an application form which included the following clauses;

> 'The inspection carried out by the Society's Valuer is not a structural survey and there may be defects which such a survey would reveal. Should you wish to arrange for a structural survey this can be undertaken by the Society's Valuer, at your own expense, at the same time as the Society's Report and Valuation is made.
>
> I understand that the Report and Valuation on the property made by the Society's Valuer is confidential and intended solely for the consideration of the Society in determining what advance (if any) may be made on the security, and that no responsibility is implied or accepted by the Society or its Valuer for either the value or the condition of the property by reason of such Inspection and Report.'

The staff valuer's report recommended a retention of £5,000 out of a total advance of £39,000 until certain specified work was done to the roof, electrical system and external decorations. No reference was made to the fact that the building spanned the river and was of very peculiar construction.

The plaintiff bought the property in May 1982 and let the second shop to a butcher. In June the butcher was using the lavatory at the rear of his shop when part of the floor collapsed. A subsequent survey indicated that the building was in a dangerous condition and in particular that the entire floor construction at the rear was beyond reclamation.

The plaintiff then claimed damages against the society on the ground that their valuer had been negligent in the preparation of his report. The judge (Mr J. Wilmers QC) was therefore required to consider just what, in practical terms, is expected of a valuer carrying out a building society inspection. He then had to go on to consider whether the disclaimer, which the plaintiff had signed, affected the situation.

On the first point the judge held that the valuer had fallen below the standard of care and skill reasonably expected in the circumstances. His opinion was:

'I bear in mind that [the valuer] was called on to do a building society valuation and not a structural survey. Nevertheless, I have no doubt that I must hold him not to have exercised reasonable skill or care. Given the nature of these premises, I do not think that it is possible to discharge properly the duties of a valuer unless he either looks under the building himself or, if he for some reason could not do so, ensured that he had the report of some other competent person who had done so. In either event, the actual state of these premises would have been revealed. I do not need to determine the depth of water at the time of inspection. Any valuer reporting on these premises, constructed as they are, must, in my view, either brave such water as he finds or find someone else to do it.'

The judge determined that the valuer and the building society, as his employers, were prima facie liable in respect of this negligent report. But he then went on to consider what effect the disclaimer had on the situation.

Significantly the plaintiff was an estate agent with offices in Wiltshire and Somerset. He had no experience in valuation or surveying, but he was aware of the nature of the disclaimer when he signed it and he had been given the opportunity to commission a more expensive report on the building – and declined to take up that offer. In the circumstances the judge held that the society, and their valuer, were protected by the disclaimer.

The judge considered the Unfair Contract Terms Act 1977 and said:

'It seems to me perfectly reasonable to allow the Building Society, in effect, to say to [the plaintiff] that if he chooses the cheaper alternative he must accept that the Society will not be responsible for the contents to him.'

This decision effectively confirmed that the society's valuer had been negligent in the preparation of his valuation, but by upholding the society's standard form of disclaimer it deprived the plaintiff of any remedy.

Watts and Another v Morrow (1991) 14 EG 111

His Honour Judge Peter Bowsher, sitting as Official Referee, found Ralph Morrow, a building surveyor, negligent in carrying out a Building Survey of Nutford Farmhouse in Dorset for Mr and Mrs Watts and an award of damages was made based upon the cost of repairs (£34,000) plus an amount for distress and inconvenience (£8,000).

Mr Morrow's normal method of preparing a report was to dictate his report (not notes for it) into a dictating machine as he walked around the property. As a consequence he had no written notes. The Judge said that this practice led to a report which was 'strong on immediate detail but excessively, and I regret to have to say negligently, weak on reflective thought'.

Further the Judge held that although a departure from the recommendations contained in the RICS practice note 'Structural Surveys of Residential Property' was not in itself an indication of negligence there had been an inappropriate departure by the defendent in the way in which he prepared his report.

As a general rule therefore surveyors undertaking Building Surveys should take written notes on site using an A4 Clipboard or similar and these notes should be retained on the file. Recommendations made by the RICS or other professional bodies should be followed. A surveyor who dictates on site without notes or who otherwise fails to follow established professional guidance is taking a serious risk.

Unfair Contract Terms Act 1977

This applies to England and Wales, but not to Scotland, and limits the extent to which a contract may attempt to exclude liability. Before a surveyor attempts to limit his liability he should confirm, by taking professional advice if necessary, that it is possible to do so having regard to the provisions of this Act. A surveyor will also have to consider if a limitation of liability is ethically appropriate having regard to the circumstances.

In Section 2 of the Act provisions are made in respect of negligence liability. The Act states:

'(1) A person (in the course of business) cannot by reference to any contract term or to any notice given to persons generally or to particular persons exclude or restrict his liability for death or personal injury resulting from negligence'.

'(2) In the case of other loss or damage, a person cannot so exclude or restrict his liability for negligence except in so far as the term or notice satisfies the requirement of reasonableness'.

In these contexts the term 'business' includes a profession.

In Section 3 the Act provides certain rules in respect of liability arising in contract, and states:

'(1) This section applies as between contracting parties where one of them deals as consumer or on the other's written standard terms of business.

'(2) As against that party, the other cannot by reference to any contract term (a) where himself in breach of contract, exclude or restrict any liability of his in respect of the breach; or (b) claim to be entitled (i) to render a contractual performance substantially different from that which was reasonably expected from him; or (ii) in respect of the whole or any part of his contractual obligation, to render no performance at all, except in so far as (in any of the cases mentioned in the above sub-section) the contract term satisfies the requirement of reasonableness'.

The 'requirement of reasonableness' means that the contract term should be fair and reasonable having regard to all the circumstances of which the parties were aware, or ought to have been aware, at the time the contract was entered into.

A distinction is made between those cases where a surveyor is acting for a client as consumer and those cases where the client is not a consumer, 'consumer' for this purpose meaning an ordinary member of the public making use of the surveyor's services in his private capacity in connection with his personal affairs. A client who is himself in business, or a company client, will not be a consumer for this purpose and as between such non-consumers and their surveyors the Act allows special bargains to be struck which can exclude any need to consider if the test of reasonableness need apply.

A member of the public buying a house is a consumer for this purpose. Whatever are the terms agreed between the surveyor and such a client they must pass the test of reasonableness. The normally-accepted and traditional practices of a profession, in respect of the nature of the inspection and the matters covered in the report, would be a reasonable basis for a contract between the parties. A substantial departure from normal or traditional surveying practices might be unable to pass a test of reasonableness.

In so far as a surveyor follows traditional and normal practices, and the recommendations of his professional institution, it would seem most unlikely that this would be held to be unreasonable.

Supply of Goods and Services Act 1982

This Act, which came into force on 4 July 1983, codifies the common law position as already described in this Chapter, and covers any contract under which a person agrees to carry out a service. The Act provides implied terms which are deemed to be incorporated into every contract for the supply of a service in that the supplier (the surveyor) will exercise reasonable care and skill, that he will carry out the service within a reasonable time and that the charge made will be a reasonable one (terms as to time and charges apply only when these matters are not covered by the contract itself). Only if the contract fixes neither the fee nor the manner of its assessment (i.e., an hourly rate for professional time) may a client challenge a charge made using the Act's provisions.

It is felt that the Act will not make any significant change in the legal liability for negligence in surveying buildings, merely that it has codified existing law by statute.

Professional indemnity insurance

As the professional's liability at law has been put to the test increasingly over the last two decades, so the insurance industry has sought to provide a constructive and effective form of protection. With the considerable growth in claims against surveyors for negligence, the demand for appropriate insurance cover has risen dramatically. Understandably, as a professional partnership does not enjoy limited liability status, the need to transfer professional risk by means of an insurance contract has become of paramount importance to all practices. The primary function of such insurance is to provide an indemnity to the assured for claims made by third parties against them arising out of professional negligence or, more importantly, alleged professional negligence in the course of the firm's business.

Insurers provide cover on what is termed a 'claims made' basis. The policy is designed to respond to claims or circumstances that might lead to claims, notified to underwriters during the currency of the policy irrespective of when the error that gave rise to the claim/circumstances was first committed. Therefore, theoretically, the insurers of a practice today could be asked to provide indemnity in respect of claims arising out of work carried out ten or more years ago. Policies of this type are known to be 'fully retroactive' and will cover the assured for claims arising out of the practice's work since the establishment of the business. Consequently, it is important to notify insurers as soon as possible of any circumstances that may give rise to a claim.

Within the term 'assured' underwriters provide indemnity in respect of work undertaken by partners, employees and those under contract of service to the practice. Furthermore, the policy should cover past partners and the estates and legal representatives of a deceased partner. Should the practice be dissolved, then it is of vital importance to arrange suitable run-off cover, as the partners will still be jointly and severally liable for work done in the past.

It is not easy to identify an adequate limit of indemnity for any given practice. Clearly, it must bear some relationship to the size of the firm and the type of business undertaken. Very large partnerships will normally purchase indemnity limits running into many millions in order to provide themselves with full protection.

Costs and expenses in defending a claim are also provided for under the policy with the consent of underwriters, and are paid in addition to the indemnity limit. Naturally, this aspect is of importance when one considers both escalating legal costs and the possibility of prolonged litigation in the event of a claim. Indeed, this time perspective should be taken into account carefully when estimating the required limit of indemnity. What might seem to be adequate cover in today's terms may well be insufficient when a particularly complex and lengthy claim notified now is eventually settled in a number of years' time.

The Royal Institution of Chartered Surveyors compulsory professional indemnity requirements dictate that cover must be on an 'each and every claim' basis. Thus, instead of providing cover of, say, £500 000 in total over the twelve months policy, underwriters will provide a limit of £500 000 in respect of each and every claim. Cover must be on a 'civil liability' basis, which provides an indemnity to the assured against claims emanating from any liability that the assured may have at law in connection with the professional business of the practice. Cover should also include fidelity guarantee, costs of restoring or replacing lost documents and a proportion of the legal costs incurred in recovering professional fees due.

Insurers always insist that the assured contribute a certain payment to each and every claim, this is known as the 'excess'. Normally this very small in relation to the limit of indemnity provided by the policy. It is felt that this device will ensure a certain degree of caution in the assured's activities. Underwriters feel it is important that the assured knows that he or she will still have a 'financial liability' to some extent in relation to a successful claim brought about by any negligence.

As with all insurance contracts the cover is governed by a set of conditions, which should always be studied carefully. All policies require the assured to provide immediate notice in writing of any claims or circumstances that may give rise to a claim, and to provide insurers with all reasonable assistance in defending such claims. As with policy conditions, policy exclusions feature in the wording, and detail those activities or types of claim that will not be covered.

It would seem that as society becomes more litigious and professionals are increasingly being asked to account for their actions, so the need for professional indemnity insurance becomes all the more essential. If the partnership is to continue its professional business without the threat of suffering unacceptably high losses as a result of human error, then appropriate steps must be taken to transfer the risk to a specialist insurance carrier.

Chapter 20

Dilapidations

The rising damp is cured now – the dry rot soaks it up.
Evil landlord to gullible tenant.

In addition to building surveys for intending purchasers and tenants, surveyors will also be asked from time to time to carry out surveys in connection with dilapidation claims, acting for either the landlord or the tenant.

Before proceeding it is essential that the surveyor understands the law involved because he, or she, will need to be able to read and interpret the lease and consider the application of certain statutes which provide some protection for tenants against claims in many situations. Although most claims are by landlords against tenants it is also possible for a tenant to have a right to claim against a landlord for breaches of repairing covenants and surveyors may occasionally be asked by a tenant to act in such a matter. The law referred to in this chapter is that applicable to England and Wales.

Claims by tenants against landlords of residential property sometimes feature in the law reports because tenants are generally legally aided or supported by housing action groups of one kind or another and many of the claims by tenants in recent years have been against local authority landlords. For surveyors in private practice the bulk of dilapidations work will involve commercial premises and the majority of claims will be by landlords.

The first thing that ought to be said is that if buildings are in poor repair and if the tenants are responsible they should do the work. There is nothing to be gained in engaging in a prolonged dispute about repairs which clearly need doing. In fact many good buildings are allowed to deteriorate due to poor maintenance and ineffective property management and if managing agents were more diligent the country's building stock would be in a better state.

This said there will undoubtedly be cases from time to time where a tenant wishes to dispute a landlord's Schedule of Dilapidations and in England and Wales there are nine principal grounds for objection which can normally be used. When acting for landlords or tenants it is worthwhile checking through these nine important matters to confirm whether, and to what extent, they may apply to the case under review.

The nine grounds for objection are as follows:

1. Defective notices

Material errors may result in a landlord's schedule being unenforceable. This will apply particularly if a landlord is serving notice under section 146 of the Law of Property Act 1925 and claiming damages or forfeiture. These notices have to be served in a particular form and the schedule must specify precisely what work is required. A vague schedule is not acceptable to the courts.

If the lease was originally granted for a term of more than 7 years and there are more than 3 years to run the notice must include certain additional information required by the Leasehold Property (Repairs) Act 1938 – an act intended to protect tenants.

These formal notices are best served by a solicitor using the schedule prepared by the surveyor. We do not recommend that surveyors serve notice of disrepair themselves claiming damages or forfeiture having regard to the legal processes which may follow if the claim is disputed.

Schedules are usually described as being either Interim Schedules or Terminal Schedules. These terms have no particular legal significance. In law a schedule may be served at any time whilst the lease is running (subject to the tenant's right to apply for relief under the Leasehold Property (Repairs) Act) or for up to 12 years after the expiry of the lease if it is by deed or 6 years after if it is not by deed (Limitation Act 1939 s.2). Any financial settlement of a terminal schedule will, however, normally be in full and final settlement of all claims so it is more important with a terminal schedule to ensure that nothing is missed. If an interim schedule is served and something is missed the landlord can usually serve a fresh schedule at a later stage.

2. Repairs may be outside the demised property

Before serving any notice on behalf of the landlord, or accepting it on behalf of the tenants, a check is advised of the description of the demised property in the lease, and the lease plan. Clearly if the repair required, or the compensation demanded, relates to matters outside the demise or outside of any common parts referred to in the lease, the tenant is not liable.

For example, in *Hatfield* v *Moss* [1988] 40 EG 112 the Court of Appeal had to consider whether the roof space on the top floor of a building divided into six flats was included in the demise of the top floor flat. The lease did not mention the roof space and a plan attached to the lease showed a roof space outside the line enclosing the demised property. The defendant had converted the roof space into a playroom and storeroom. See also *Straudley Investments Ltd* v *Barpress Ltd* [1987] 282 EG 1224 where the issue turned on the rights to construct a fire escape and ventilation vents on and against the roof of the demised property.

In *Hatfield* the Court of Appeal held that a plan which is for identification only cannot control the parcels clause in the lease which describes the extent of the demise, however if the parcels clause is unclear then it was appropriate to look at the plan for guidance regarding what may be included or excluded within the demised property. Lease plans are normally described as being attached for identification purposes only, and not all leases have plans. So it is the written description which one must consider first and if this is clear then whatever is shown on the plan does not change the position. Only if the written description is unclear can one resort to using the plan and any surrounding circumstances to settle the issue.

3. The lease may not be on full-repairing terms

A careful reading of the lease may show that the Repairing Covenants do not include the repair required: in other words the lease may be something less than a full-repairing lease.

Covenants that refer to 'Good and Substantial Repair' will normally be effective as full-repairing covenants. Covenants that refer to 'Good and Tenantable Repair' require a standard of repair suitable for a tenant carrying on the particular kinds of businesses for which the property is leased which, in the case of some types of industrial or retail premises, may not be as high as 'Substantial' repair. Some covenants refer only to a tenant's liability for interior or exterior decorations and may be silent on the question of liability for major repairs, or may require that the landlord repairs the 'exterior and structure'.

If in doubt it is best to refer to the solicitor for an interpretation of any covenants that appear doubtful. If a schedule is served on behalf of the landlord it could rebound to his disadvantage if it includes matters which are found, on examination, to be landlord's repairs.

4. Section 18 (1) of the Landlord and Tenant Act 1927 may apply

Section 18 (1) of the Landlord and Tenant Act 1927 provides that a ceiling is set over any landlord's claim for damages which cannot exceed the loss in the value of his interest in the property (diminution in the value of his reversion). For example, if the landlord can be shown to be proposing to alter or demolish a property its state of repair may be irrelevant and the tenant may happily allow it to deteriorate, notwithstanding that he may have a full-repairing lease.

The section reads as follows: 'Damages for the breach of a covenant or agreement to keep or put premises in repair during the currency of a lease or to leave or put premises in repair at the termination of a lease, whether such agreement is expressed or implied and whether general or specific, shall in no case exceed the amount (if any) by which the value of the reversion (whether immediate or not) in the premises is diminished owing to the breach of such covenant or agreement as aforesaid; and in particular no damages shall be recovered for a breach of any such covenant or agreement to leave or put premises in repair at the termination of a lease, if it be shown that the premises, in whatever state of repair they might be, would at or shortly after the termination of the tenancy have been pulled down, or such structural alterations made therein as would render valueless the repairs covered by the covenant or agreement.'

With some types of property the state of repair of particular items may be unimportant in terms of letting value. For example, shop premises are invariably refitted by the incoming tenant, often at great expense to suit a particular corporate image, so the state of the interior and shopfront left by the outgoing tenant may not be important.

In *Mather* v *Barclays Bank Plc* [1987] 2 EGLR 254 the landlord claimed against an outgoing tenant for substantial dilapidations, however new tenants were found (a building society) who undertook not only the repairs but also substantial improvements. The investment value of the landlord's reversion following the letting to the new tenant, capitalising the rent at a 7% yield, produced a valuation well above that for the property repaired but unimproved. The court held that the landlord's claim failed under the first limb of Section 18 (1) of the Landlord and

Tenant Act 1927 as the value of the reversion had not – as it turned out – been diminished by the breach of covenant.

Outgoing tenants would be wise to delay settlement of a claim for as long as possible to see what happens to the property on reletting.

5. Internal decorations may be excluded

It may come as a surprise to many tenants to know that a landlord cannot automatically require internal decorations to be carried out during the term of a lease and an interim schedule which includes internal decorations can be challenged by an application to the court for relief. (A terminal schedule at the end of the lease may quite properly include internal decorations however.) This is provided for in Section 147 of the 1925 Law of Property Act. Note however that decorations which are required to prevent deterioration to building components or ensure compliance with covenants to repair or ensure cleanliness must still be undertaken.

The section provides as follows:
(1) After a notice is served on a lessee relating to the internal decorative repairs to a house or other building, he may apply to the court for relief, and if, having regard to all the circumstances of the case (including in particular the length of the lessee's term or interest remaining unexpired), the court is satisfied that the notice is unreasonable, it may, by order, wholly or partially relieve the lessee from liability for such repairs.
(2) This section does not apply:
    (i)   Where the liability arises under an express covenant or agreement to put the property in a decorative state of repair and the covenant or agreement has never been performed;
    (ii) to any matter necessary or proper –
        (a) for putting or keeping the property in a sanitary condition, or
        (b) for the maintenance or preservation of the structure;
    (iii) to any statutory liability to keep a house in all respects reasonably fit for human habitation;
    (iv) to any covenant or stipulation to yield up the house or other building in a specified state of repair at the end of the term.

6. Tenant may not be liable to renew, as opposed to repair

This is a more arguable area since the protection provided for the tenant is not a statutory protection but the protection of the common law and subject to legal precedent. In general the courts have held as a matter of principle that a tenant is not liable to hand back anything which is improved to the point where it is wholly different from that originally demised. Repair will include some replacement of parts in many cases but should not extend to the point where it constitutes renewal of the whole. (A lease may, of course, provide for renewal rather than repair but a sensible tenant should avoid any such commitment unless the implications are fully understood and acceptable.)

The extent of a repairing covenant can have implications in other areas of Landlord and Tenant Law such as rent review. In *Norwich Union Life Assurance Society* v *British Railways Board* [1987] 283 EG 846 a repairing covenant in a 150-year lease with provisions of rent reviews at 21-year intervals required the tenant to 'keep the demised premises in good and substantial repair and condition and when necessary to rebuild, reconstruct or replace the same'. An arbitrator dealing with a rent review dispute considered that this placed on the tenant a more onerous obligation than the usual covenant to keep the premises in good and substantial repair and he accordingly made a downward adjustment of 27.5% in the rent to reflect the difference.

The court upheld the arbitrator's view and the appeal against his award was dismissed. The judgement recorded that the language of the repairing covenant included two separate obligations; first, that of repair and secondly that of rebuilding, reconstructing or replacing the entire premises.

In a normal case what actually constitutes repair or renewal must depend on the circumstances. Underpinning of part of a building has been held to be a repair notwithstanding that it is also an improvement to remedy an inherent defect (see *Rich Investments Ltd* v *Camgate Litho Ltd* [1988] EGCS 132).

Eventually a roof covering needs renewal but in *Murray* v *Birmingham City Council* [1987] 283 EG 962 the court held that the roof of a terraced house built in 1908 had not yet reached the stage where complete renewal of the roof was the only option and piecemeal repairs were still a valid option for the landlord. In this case the tenant on a weekly tenancy wanted a new roof and the local authority landlords wanted to carry on undertaking patching repairs to the existing roof.

7. The Leasehold Property (Repairs) Act 1938 may apply

In any case where a lease was granted for a term of more than 7 years and there are more than 3 years to run the landlord may not proceed with service of notice of disrepair and schedule of dilapidations without the leave of the court.

A preliminary court hearing is required and the burden of proof is on the landlord to show that the repairs are required on one of five grounds laid down in the Act, the most important of which is that the repair is required to prevent substantial diminution in the value of the landlord's interest. As Lord Denning MR said (*Sidnell* v *Wilson* 1966): 'That Act was passed shortly before the war because of a great mischief prevalent at that time. Unscrupulous people used to buy up the reversion of leases, and then bring pressure to bear on tenants by an exaggerated list of dilapidations.'

At the preliminary court hearing surveyors will often be called to give evidence on the matter of the state of repair of the premises and the significance of dilapidations as they may affect the value.

The five grounds are:

(a) That the immediate remedying of the breach in question is requisite for preventing substantial diminution in the value of the reversion, or that the value thereof has been substantially diminished by the breach;

(b) that the immediate remedying of the breach is required for giving effect in relation to the premises to the purposes of any enactment, or of any byelaw or other provision having an effect under an enactment, or for giving effect to any order of a court or requirement of any authority under any enactment or any such byelaw or other provision as aforesaid;

(c) in a case in which the tenant is not in occupation of the whole of the premises as respects which the covenant or agreement is proposed to be enforced, that the immediate remedying of the breach is required in the interests of the occupier of those premises or part thereof;

(d) that the breach can be immediately remedied at an expense that is relatively small in comparison with the much greater expense that would probably be occasioned by postponement of the necessary work; or

(e) special circumstances which in the opinion of the court render it just and equitable that lease should be given.

Note: These paragraphs should be read as if 'or' came between each of them so that the landlord need only satisfy the court on one ground (see *Phillips* v *Price* [1959] Ch. 181).

For practical purposes the effect of this Act is to prevent a landlord serving a notice of disrepair and schedule until the last 3 years of the term except in cases of fairly extreme disrepair. For managing agents perhaps it would be sensible to make a diary note to inspect all properties where the original leases were for 7 years or more at the time when they have 3 years to run. If there are going to be arguments about dilapidations there is then plenty of time to inspect, serve notice and follow the matter up before the lease actually expires.

8. A schedule of condition may exist

When a lease is granted it is often a good idea for landlord and tenant to agree a schedule of condition, describing the property fully, as evidence of its state of repair and then have copies attached to the lease and counterpart lease. This is particularly important from the tenant's point of view if the property is old and in poor repair. The lease may then provide that, notwithstanding the provisions of any repairing covenants, the tenant will not be liable to hand the property back to the landlord on termination in any better condition than is evidenced by the schedule of condition.

Surveyors acting for tenants in receipt of dilapidations claims would be advised to check the lease to see if there is any reference to a schedule of condition and if so to locate a copy of the schedule.

The schedule normally consists of an item-by-item description of the property and its various components, frequently with photographs.

9. The landlord may be liable for implied repairing covenants

Sections 11–16 of the Landlord and Tenant Act 1985 (formerly Sections 32–33 of the Housing Act 1961) provide that when a residential tenancy is granted for less than 7 years the landlord always remains liable for certain matters. It is not possible to contract out of this liability, except by granting a residential lease of more than 7 years.

The implied covenants are:

(a) to keep in repair the structure and exterior of the dwelling-house (including drains, gutters and external pipes),

(b) to keep in repair and proper working order the installations in the dwelling-house for the supply of water, gas and electricity and for sanitation (including basins, sinks, baths and sanitary conveniences but not other fixtures, fittings and appliances for making use of the supply of water, gas or electricity), and

(c) to keep in repair and proper working order the installations in the dwelling-house for space heating and heating water.

It should be noted that Section 11 of the 1985 Act is amended by Section 116 of the 1988 Housing Act in respect of leases granted on or after 15 January 1989. Under these provisions the implied covenants extend to 'any part of the building in which the lessor has an estate or interest' if the 'Disrepair . . . is such as to affect the lessee's enjoyment of the dwelling-house or of any common parts, as defined in Section 60 (1) of the Landlord and Tenant Act 1987, which the lessee, as such, is entitled to use'. This extension of the lessor's liability is intended to cover situations which may well arise in blocks of flats or buildings in multiple occupation.

Taken together these provisions are intended to ensure that tenants in short term residential lettings can insist on certain basic minimal standards of maintenance and the provision of basic sanitation including hot water and central heating where these exist at the commencement of the letting. These provisions do not apply to lettings of commercial premises.

The inspection

The foregoing is a very condensed summary of the legal position as it applies to notices of disrepair and the negotiations which may take place between surveyors acting for landlords and tenants. Most claims are settled amicably by negotiation against this legal background.

Since this is not intended as a legal text book I have not attempted to cover the legal aspects in any detail and readers interested in delving more deeply into the complexity of dilapidations claims will be able to find much of interest in the various legal text books.

From the surveyor's point of view these matters generally arise initially with a request from a landlord to inspect a property and advise whether a schedule is required, or from a tenant who is in receipt of an unwanted schedule. Often an informal list of repairs is prepared first and sent, with an initial letter, to the tenant pointing out the various matters that need attention. This informal letter may be followed by a formal notice of disrepair and schedule later if the repairs are not attended to.

The inspection procedure will be similar to that required for a building survey with a need for particular thoroughness if a terminal schedule is involved since once a full and final settlement is reached the landlord will not be able to claim for anything which may have been missed (unless, that is, he decides to claim against his hapless surveyor!).

The notes will be taken in tabular form covering all elements of the construction internally and externally and listing all work needed to comply with the repairing covenants.

If a case is likely to go before the court then it is normal for the schedule to be prepared in the form of a 'Scott Schedule' with columns for each item, notes from any schedule of condition, details of the alleged dilapidations, the landlord's claim, the tenant's response, the plaintiff's observation, the defendant's observation and a final column for the judge's notes.

During any hearing the judge may then move briskly through the schedule dealing with each item in turn and noting his award of damage (if any).

If the tenant is unwilling or unable to do the work the claim will be expressed financially and it will be necessary to cost out each item, so some measurements will generally need to be taken during the inspection in order to calculate the cost of some repairs, expressed as a price per square metre or the equivalent.

In many cases Section 18 (1) of the 1927 Landlord and Tenant Act will be invoked by the tenant, either to repudiate the claim altogether or to justify a reduction in the total claim. If it is clear that the landlord proposes to demolish or reconstruct the property in such a way that its condition is immaterial to its value to him then any claim will be unsupportable and there would seem to be little point in spending valuable time preparing a schedule in the first place.

If the tenant claims that the landlord has such intentions, but the landlord denies it – or claims to have had a change of mind – it would be sensible for the solicitors to resolve this preliminary point before instructing surveyors.

Even if the landlord has no such proposals there may be arguments about diminution in value. In such cases it is best to complete the schedule first and calculate the cost of repairs; then to look at the total arrived at in relation to the value of the property as it is and as it would be with the repairs done.

In some cases there will be no difference – the cost of repair will be the same thing as the difference in value. In other cases some repairs will be judged unimportant in relation to the value of the property.

Calculation of the amount by which the reversion has been reduced in value is undertaken on a simple 'before and after' basis taking the value of the landlord's interest (which may not necessarily be freehold) as it stands and then as it would be had no breaches occurred.

The survey for dilapidations purposes should therefore also include measurements and other details of the property because the open market value of the landlord's interest will have to be assessed. With retail premises, for example, the location from a trading point of view will be very important and probably of more significance to its value than matters such as shop fittings which have a traditionally short life. Most incoming tenants of retail units will budget for refitting the interior anyway and the actual condition of the interior may well be irrelevant to letting value (and hence the value of the landlord's reversion).

It is worth noting at this stage that the statutory protection provided for tenants does not apply the other way. If a landlord has a repairing liability and is in breach then the tenant may sue for damages. The tenant's claim will succeed or fail simply by the application of the law of contract to the particular facts as evidenced by the surveyors to each side as expert witnesses on the issues of liability, cost of repairs and incidental losses which the tenant may have suffered (e.g. from leaking roofs).

Statutory dilapidations

In addition to the liability to repair arising under the law of contract and the Housing Act there is a variety of statutes which may make property owners, occupiers or building contractors liable in various circumstances for repairs or other works to buildings in order to comply with the law. Some of the more important statutes are as follows:

Fire Precautions Act 1971 (S1 and 2)

This designates premises for which fire certificates are required. The condition of fire escapes, alarm and sprinkler systems and the means of preventing spread of fire are all important features in any dilapidations survey. From 1 April 1989 all factory, shop, office or railway premises where more than 20 people work, or where more than 10 people work at any one time on a floor other than the ground floor, must have a fire certificate. Even if a fire certificate is not required, because fewer than the stipulated number of people work there, it is still necessary to provide the means of escape and the means of fighting fire. The local fire service can advise individual property owners on the action needed to comply.

Defective Premises Act 1972

This places a duty on a builder to build properly (S1) – a duty which is not discharged by any subsequent disposal of the building (S3).

Offices, Shops and Railway Premises Act 1963

Specifies minimum requirements for space, temperature, ventilation, lighting and sanitary conveniences. Note especially the requirements for adequate lavatory and washing facilities, including hot water, for both sexes.

Building Regulations

These include provisions for enforcement in respect of new buildings or building alterations. During a dilapidations survey it is necessary to look out for unauthorised alterations which may have been undertaken without either landlords' or local authority consent.

Local Acts

At one time there were hundreds of local Acts of Parliament which applied to particular boroughs, cities or counties. Most of these have been repealed but some remain in force in mainly urban areas affecting such diverse matters as access for fire fighting, storage of inflammable liquids, temporary structures, dust from buildings and the fixing of signs. If in doubt it is wise to check with the building control department of the local authority whether any local Act applies.

London Building Acts (Amendment) Act 1938

Originally a local Act covering the special code for party wall procedures within inner London, intended to facilitate the development of urban sites where a need may arise to build into or on existing party walls, or repair party walls. The Act provides a procedure for resolving disputes. The provisions were found to work well in inner London and were extended in 1996 to apply to the whole of England and Wales.

Factories Acts

Certain minimum standards are provided for similar to those that apply under the Offices, Shops and Railway Premises Act.

Conclusion

When dealing with dilapidations claims it is wise to keep in mind the importance of maintaining good relations between landlords and sitting tenants – they need one another. The tenant needs the premises – the landlord needs the income. Also the fact that dilapidations issues often arise at the same time as other negotiations for lease renewal, assignment, change-of-use or rent review. Frequently dilapidations matters are settled as part of an on-going property management strategy. Careful study of the law involved in all these areas is recommended because a dilapidations claim against a sitting tenant is rarely a matter that can be considered entirely in isolation.

Chapter 21

Conservation and the surveyor

Another giant glass stump better suited to downtown Chicago than the City of London – a monstrous carbuncle on the face of a much loved and elegant friend. HRH The Prince of Wales referring to proposed developments adjoining St Pauls, at the RIBA 150th Anniversary Banquet in 1984.

We are concerned in this final chapter with building surveys undertaken not for prospective buyers, mortgage lenders, landlords or tenants, but those undertaken for existing property owners needing advice on the conservation of their property. Conservation is now a major national and world issue. We must avoid the wasteful use of the world's resources including the fine buildings passed down to us by previous generations. We must be prudent and economical in our use of energy and building materials. I make no apologies for devoting a whole chapter to this issue.

Conservation principles

Surveyors can assist conservation in its widest sense if they aim to further the following objectives when undertaking their day to day work:

1. To design and construct buildings with a long life span requiring minimal maintenance and capable of adaption to changing needs and the requirements of occupiers.
2. Avoid the wasteful use of materials and energy in the construction process and subsequent life cycle maintenance.
3. To design buildings and modify buildings so as to make them easily repairable with convenient means of access to structural components and services.
4. To plan the maintenance of existing buildings over their life cycle and avoid damage by neglect.
5. To use sympathetic materials and techniques when maintaining existing historic buildings.

Practical conservation techniques are applied in the main to buildings which are of historic or aesthetic importance including most ecclesiastical structures, listed buildings of all types, tithe barns, country houses and even art deco and post-war property where the design or construction is of special interest. Conservation is increasingly important now with the growth of Conservation Areas designated under planning legislation, and Georgian, Victorian and Edwardian terraces are enjoying renewed popularity in many of our towns and cities.

As conservation generally requires a greater degree of maintenance of a continuing nature than would be required for a building of less importance, a great deal of thought and planning should have been employed in ensuring the most effective use of the physical and financial resources available. If the necessary finances and technical expertise have not been available in the past then the surveyor must be extremely wary and meticulous in carrying out his, or her, inspection.

Preparations for a survey

The agencies causing deterioration and subsequent decay in an older building can generally be attributed to the weather, vandalism and misuse, the forces of gravity, chemical attack and biological attack from fungus and insects. People are often responsible for much of the damage.

Before carrying out an inspection, and prior to preparing a building survey, one must therefore consider the original design, its structural advantages and weaknesses, undue wear or damage caused by human agencies and the natural elements and their impact. Sadly unattended older buildings require special attention and security since they were not designed or constructed to resist crowbar or aerosol.

Thankfully in this country we are fortunate in the scarcity of natural disasters such as earthquake or flood and no special precautions are needed against such eventualities but often too scant regard is paid to the effects of gale force winds and snow against which some precautions should have been taken by those responsible for the upkeep of the building.

Before dealing in detail with any of these matters, it must be remembered that during the process of any conservation work to an important building there should have been no interference with any historical evidence which may be in the building or on the site. Before commencing such work the owners should have taken particular care in compiling a fully documented record to include materials and designs used in the construction. Any such information already available should be closely studied and logged by the surveyor before he, or she, begins the survey.

Before commencing any work to a 'special' building, preparing specifications or seeking tenders a full condition survey should therefore be made and a comprehensive report on the structure prepared. This will then indicate how the materials and methods of construction used in the structure have performed in the past and facilitate an evaluation of the likely future performance of that building.

From such a report, in consultation with the owners, the 'degree of intervention' required to carry out the conservation work may be assessed, taking one of the following listed in rising order of cost and complexity:

(1) Simple prevention of further deterioration
(2) Preservation in existing condition
(3) Structural consolidation
(4) Restoration of structure, fittings and finishes
(5) Rehabilitation including possible improvement items
(6) Reproduction – imitative or by substitution
(7) Reconstruction in whole or in part

Before starting the survey it is useful for the surveyor to know in broad terms what level of expenditure and degree of intervention the owner has in mind (most owners will naturally want a Rolls-Royce job for a Morris Minor price). At this point it is also important to establish whether the works, or any part of the works, are to be grant-aided from charitable or public funds, since the standards and requirements of the organisation providing the funding will have to be met.

Conservation work generally demands the services of highly skilled and experienced craftsmen (and craftswomen) – stonemasons, for example – who have a long lifetime of experience of traditional methods and materials. In many cases good stonemasons, carpenters, joiners and even plumbers must be artists just as much as tradesmen. The choice of labour for this type of work is particularly vital.

Obviously it is important that the surveyor does not recommend solutions which are impractical because the skilled trades needed are simply not available in the locality.

To summarise these preliminaries which it may be possible for the surveyor to undertake prior to the condition survey:

(1) Assess environmental, human and natural agencies causing deterioration.
(2) Obtain available documentation including any historical evidence.
(3) Establish the degree of intervention the owner has in mind and resources available.
(4) Consider the type and availability of the craftsmanship needed.

Techniques, structures and materials

Many older structures to which conservation techniques are applied are less sophisticated and more massive than modern structures which offer equivalent accommodation. Now, in the age of the computer and with the Building Regulations and Codes of Practice offering satisfactory minimum standards, buildings tend to be 'built-down' to these calculated minimum standards and so modern construction often appears to be surprisingly fragile although in fact generally adequate for the forces involved.

The reasons for the more massive material content of the older buildings are that weaker materials were used and 'rule of thumb' was the order of the day, the tendency being to ensure stability by over-designing with extravagant use of material resources.

Another reason for the provision of strength by mass and sheer size was the intended permanence of structures. Short life public buildings were probably unknown in Norman times and during the Middle Ages. Even the sophisticated Romans and Greeks built with permanence in mind and we can still benefit from the results. Compare, for example, Wells and Canterbury Cathedrals with the Roman Catholic Cathedrals of Liverpool and Bristol; which are the more massive and which are likely to have the greater permanence? But in mitigation of modern design, which are likely to be the more maintenance-intensive, original construction defects apart?

The first thoughts of a surveyor on considering the performance of an older building would be directed towards the loadings on the different parts of the structure and, in particular, the foundations. Modern road traffic – increasing in both volume and weight – uses roads which were never designed for the proper transmission of the consequent loadings and the ground vibrations thus caused by such traffic can cause foundation displacement with resultant movements in the upper structure.

Rapid and irregularly applied loads have a far more serious effect on a structure than dead loads or gradually applied loading, but by far the worst damage is caused by regular, rhythmic, repetitive loads which, coupled with the natural resonance of the structure, cause eventual collapse. The tragic loss of the Tacoma Narrows bridge in north-west USA was an example of such a phenomenon and the considerable damage caused to the undercroft and structure of York Minster demonstrates the effect of modern traffic vibrations.

In addition to distortion of the structure caused by the stress and load problems mentioned above, settlement and distortion may have been caused by the action of materials used. These can be broadly divided into defects which occur in the first phase of a building's life – say 25 years from completion – and those which take place at a later date or even have a cumulative effect over the years.

Deterioration or initial shrinkage of materials such as mortar may be responsible for initial deformation of the structure but longer term effects are caused by weathering, internal or external severe temperature changes, the actions of salts (e.g. sulphate attack or efflorescence), rising damp and changing moisture content in the materials used. Modern buildings use more elastic pointing materials such as mastics and they incorporate movement joints at appropriate intervals if they are well designed. Moisture rising from the site is now well controlled by damp proof courses and membranes.

Cracking in materials is a regularly occuring phenomenon in all old buildings. Superficial cracking occurs in applied finishes such as rendering and plaster but the more serious faults are caused by the cracking of the structure itself. For example, cracks caused by settlement in the structure or subsidence which may be manifested as raking cracks in a brick wall, initially following the lines of the horizontals and perpends to the mortar joints but eventually cracking the bricks themselves. Brittle stone is often strong in compression but weak in tension, such cracks appearing parallel to the compression force where weakening of the structure has consequently occurred.

Poor quality restoration was apparent in a 'listed' high-rise block seen in the mid-west of the USA which is entirely clad in glazed clay faience blocks and had recently undergone renovation. The renovation seemed to have been applied principally to surface cleaning of the glazed blocks and was largely cosmetic in nature. Many cracks in the glazed surfaces of the blocks with associated crazing of the glaze were evident and it is inevitable that the building envelope has become less than watertight with consequent corrosion of the steel structural frame beneath, exacerbating the cycle of further cracking, more water penetration and further corrosion of the steel frame. It is fortunate indeed that buildings in the USA are generally regarded as expendable with a very limited life.

Compare this with the Inca structures at, for example, Cuzco and Macchu Picchu in Peru. These were built of huge blocks of granite, interlocking, completely devoid of mortar with walls having an out-of-plumb inward batter and it is impossible even to insert a penknife in the joints between the blocks. These buildings were, as a result, so strong that they were virtually earthquake resistant and very permanent in nature.

Although in this country we do not have to design against earthquakes and even minor earth tremors are a rare occurrence, damage to the structure may be caused by excessive water abstraction causing a drying out of the subsoil, on occasion by the collapse of obsolete sewers, damaged water mains causing subterranean cavitation or even – as in Northwich, Cheshire – older houses disappearing into cavities formed by subterranean brine extraction. We are likely to hear increasingly in the future about changes in the water tables causing damage to many buildings, especially in London, as increasing use is made of bore holes as a means of supplying a rising demand for water.

To sum up, the environment in all its aspects must be closely studied to assess the effect of all these factors on the building before commencing the survey itself.

Local techniques of construction must be investigated as it is most important that the reason why a building has been designed and built in a particular way out of certain materials should be appreciated before reporting on the structural condition. Some rural areas in the UK have a great deal of cob and thatch. The thatched roofs have no eaves gutters, thus dispensing with the rainwater disposal problem; the eaves project further than usual and there is a slow 'run-off' from the semi-absorbent thatch. The 2 to 3 feet thick cob walls are plastered both sides to prevent ingress and egress of damp, the cob (straw, clay and earth) core forming a moist cohesive 'sandwich'. On no account should a damp proof course be recommended as the consequent drying out could leave one with a pile of red dust instead of a wall! Rising damp is, however, controlled by building the first metre of a cob wall in a semi-absorbent stone such as local sandstone and a very effective method of construction this is, provided the integrity of the internal and external plastering is preserved.

In the Lake District stone walls are laid in thick slate tilted so that the joints slope outwards to prevent damp penetration. Unfortunately in Castle Drogo, Devon – completed around 1930 – the walls are battered and the mortar joints slope inwards resulting in excessive damp penetration. The face of each block should have been battered and the joints made level or sloping outwards. Was this lack of thought or inadequate supervision?

Space does not permit a description of all the many regional variations in the use of materials and forms of construction but the examples given above should be sufficient to show that, before embarking on the survey of an old building, an intensive knowledge of local historical methods of construction and the use of materials is required.

Many regions in this country also have their unique problems covering insect infestation and fungoid infection. The House Longhorn Beetle is localised in the southern home counties and especially the Weybridge area of Surrey. Furniture beetle infestation appears to be more prevalent in the south and south-west than in the north, whereas dry rot seems to spread more rapidly in the north of England than in the south. As older buildings are more likely to suffer structural damage from these causes than are the newer buildings where preventive measures have often been taken, the surveyor must study the potential impact of these factors on the subject property.

When carrying out repairs, traditional materials compatible with the locality should have been used wherever possible but in some cases modern substitutes will have been the only remedy available – for example, stone dust/epoxy resin in stonework repairs. Such substitutes should only have been used as a very last (and extreme) resort where no other means were available. One particular fault often encountered is the repair of gauged brick arches over windows and door openings, often clumsily executed with inferior bricks and sand/cement mortar where soft, sandy 'rubbers' and lime mortar should have been used. As a general rule the use of

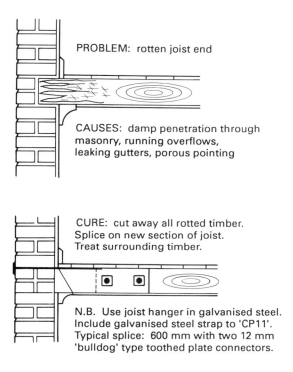

PROBLEM: rotten joist end

CAUSES: damp penetration through masonry, running overflows, leaking gutters, porous pointing

CURE: cut away all rotted timber. Splice on new section of joist. Treat surrounding timber.

N.B. Use joist hanger in galvanised steel. Include galvanised steel strap to 'CP11'. Typical splice: 600 mm with two 12 mm 'bulldog' type toothed plate connectors.

Figure 21.1 Conserving floors

Portland Cement should have been avoided wherever possible in old buildings as it shrinks, is too strong for the surrounding materials and may produce salts which can cause damage.

Planned maintenance and the surveyor

There are two approaches to maintenance: to plan for it or to do the work only when absolutely necessary on an *ad hoc* basis. Effective and regular planned maintenance work should always be a major factor in the conservation process. It is a necessary feature which is the only logical way to ensure the retention of a historic or unique building so that it remains in a usable condition and does not deteriorate.

Planned inspection to a routine can be of use to surveyors in the case of larger, older and unique buildings. It can provide us with invaluable feed-back of performance information over a very long period of time, often with materials and forms of construction which are no longer in regular use and not otherwise understood.

Having a planned maintenance system for a building or group of buildings also means that management will sometimes maintain a small, directly employed labour force of specialist tradesmen selected from amongst the older and more

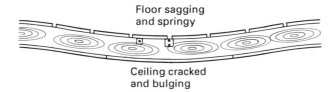

Floor sagging
and springy

Ceiling cracked
and bulging

Excessive notching-out of old floor joists. Ideally joists
should be drilled. If notching is unavoidable it should
be limited to 1/8th the joist depth. Victorian and earlier
construction often features undersized joists in any event.

CURE: slight deflection – insert new joists/additional
joists. Serious deflection – renew whole floor

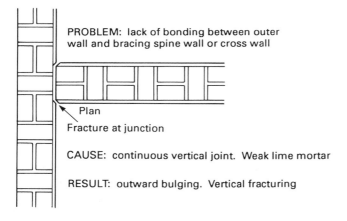

PROBLEM: lack of bonding between outer
wall and bracing spine wall or cross wall

Plan

Fracture at junction

CAUSE: continuous vertical joint. Weak lime mortar

RESULT: outward bulging. Vertical fracturing

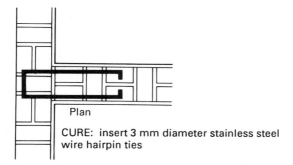

Plan

CURE: insert 3 mm diameter stainless steel
wire hairpin ties

Figure 21.2 Conserving walls

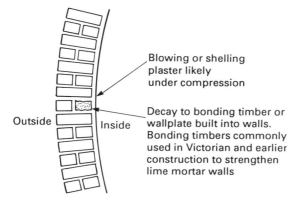

Bulging wall with timber
built into brickwork.

CURE: walls seriously out-of-plumb are better
rebuilt reusing bricks when these are sound

Figure 21.2 Continued

experienced craftsmen in the locality, together with a few younger apprentices. Such a system ensures that traditional skills in repair, restoration and renovation are handed down from generation to generation and keeps the continuity in knowledge and techniques alive and is more likely to ensure that the building is kept up to a good standard of repair.

Planned maintenance also requires a reporting system. Neglect of an old building in partial or constant use can be mitigated by briefing staff, caretakers and, in particular, cleaners to ensure that they promptly report telltale signs of want of maintenance. It is often said that the best way of ensuring that no defects occur in one's car is to clean it inside and out regularly by hand, instead of just putting it through the car wash, so that incipient rust and other defects are detected and preventive measures taken in good time. The same principle can apply to a building and very little training is needed to detect damp penetration, broken glass, leaking overflows and gutters, damaged fittings and the like. A chat with the operating staff can often provide the surveyor with invaluable performance information and direct his or her attention to those parts of the property which require a closer inspection.

Another highly valid reason for using all eyes, hands (and sometimes noses) available from the regular staff in a historic or valued building is that such properties are not, as most modern buildings, constructed to last for a limited life but are expected to be of a permanent nature. Any fungoid infection, insect infestation or damp penetration is very likely to have been noticed by operating staff and this information must be acted upon in order to avoid the onset of cumulative damage.

Great care is required in suggesting remedial measures. It is best if possible to ensure that the same methods and materials are recommended as were used in the original construction. The introduction of new 'foreign' materials can sometimes have a serious and even catastrophic effect. Some caution should be exercised when specifying timber treatment having regard, on the one hand, to the need to control wood borers and fungal attack and, on the other hand, the importance of limiting the amount of chemicals used which could have damaging long term effects on the health of operatives, occupiers and visitors. Vapour barriers must be used with care to prevent undue moisture concentration in sensitive places. The juxtaposition of materials with different characteristics must be carefully considered, e.g. hard cement mortar should not be used with soft stone or brick, nor untreated softwood fixed to a damp wall.

Vapour barriers are required on the warm side of insulation, but not on the cold side. On the cold side there must be adequate ventilation. Failure to observe these rules when insulating roof voids will result in condensation problems which can give rise to rot and other timber defects. If a roof is recovered in new tiles or slates on battens and sarking felt ventilation must have been included in the specification. Without sarking the void is probably well ventilated but when sarking is added the void is effectively sealed. With insulation added to the ceilings it is sealed and cold. Cut a hole in the ceiling for a service duct or down light – especially over a kitchen or bathroom – and warm damp air is passing through into the cold void, creating a major problem of condensation on the roof timbers. Hence the importance of considering each repair or improvement to a building as part of the total and not in isolation. During the survey inspection it is imperative that any signs of condensation must be thoroughly investigated, the causes traced and remedial measures suggested.

Plumbing, electrical and heating services all have a limited life and require periodic renewal over the life span of an ancient building. They therefore need to be accessible and renewable without disturbing the historic structure which means in most cases running cable and pipework on surfaces or within accessible ductwork. Some thought needs to be given to the conflicting requirements of accessibility on the one hand and the desirability of hiding unsightly pipes and cables on the other hand as concealed services do limit the scope and value of an inspection in the case of older buildings.

Although some abandoned buildings and ruins need merely to be kept in both a wind and watertight and structurally safe condition, most buildings to which a conservation policy has applied have received more sophisticated treatment in the form of a quinquennial plan covering major repair, pointing and decoration with regular cleaning out of gutters, gullies, drains, valleys and parapet gutters in particular – say twice yearly. In addition regular routine inspections of security and fire alarm systems, smoke and heat detectors and emergency lighting together with appropriate testing is needed in any important building.

In the case of lofty buildings such as churches, lightning conductors should be installed, inspected, tested and certified – preferably annually. Any certificates should be seen during the course of the survey otherwise an oversight on the part of the surveyor could result in a professional indemnity claim. Remember that it was lightning that caused the disastrous fire to the South Transept of York Minster.

When considering all the foregoing factors the owners' or occupiers' objective should be 'keep, restore or improve every facility in every part of a building, its services and surroundings to an acceptable standard and to sustain the utility and value of the facility'. The 'acceptable standard' for conservation purposes is often dependent upon the amount of finance available as the type of building involved is often administered by a trust, charity, a single owner or is an ecclesiastical building – in short, a 'non-productive' building in that it does not have an income-producing or profit-making function associated with it. This limited maintenance funding in such cases means that very careful pre-planning of resources available must have been carried out including a strong element of preventive maintenance to guard against expensive, catastrophic failures as well as a strict system of priorities. This should have ensured that, for example, structural and roof repairs as well as external painting and gutter maintenance have had priority over internal decorations and work of a purely cosmetic nature.

Maintenance planning for buildings can be assisted using computer-supported data bases. Surveyors dealing with maintenance planning and life-cycle costing invariably use a computer data base nowadays to store all the relevant information. Any such records of past maintenance work giving details of tasks and dates would be invaluable to the surveyor in assessing the performance of a building.

Many of the cities and towns in developed countries may now be regarded as fully built. The infrastructures are already in place and attention is shifting from construction to conservation with greater emphasis on the economical use of resources and avoidance of waste.

The initial condition survey, including the assembled dimensions and floor plans, is an essential first step in setting up a computer based maintenance planning system. A well organised building survey can form the basis from which decisions can be made regarding the future maintenance programme for the structure in question.

Access

Access for survey purposes is often a problem in older buildings and, for example, tower scaffolding is usually necessary to inspect and repair the high-level interiors of cathedrals, churches and halls with consequent extra expense. The provision of walkways and access hatches in roof voids, permanent access and walkways to high level roofs, hatches to inspect and repair sub-floor areas as well as ducts and accessible trunking for services will all help to reduce future costs and ensure that no part of the structure remains neglected. The more inaccessible the area of a building to be maintained or surveyed (e.g. a church spire or a high steeply-pitched roof) the more essential it is that any works should be carried out with a longer-term cycle in mind, even if the cost is greater at the time. One would not wish to have to scaffold the spire of Salisbury Cathedral, for example, more than once in a lifetime at the very most! Perhaps 80 to 100 years repair-free service would be the figure one would wish to achieve in such a case and a full inspection should be made every time works are carried out making maximum possible use of the access provided.

Further information

Buildings to which a conservation policy applies demand special treatment in the matters which are to be covered by a building survey and the surveyor will need to undertake continuous research and study in order to be able to deal adequately with the problems encountered.

I have not attempted with this volume to write a text book on building construction *per se*; there are many excellent books available on construction techniques and materials, some of which are mentioned in the Bibliography. Much useful information can be obtained from bodies such as the Ecclesiastical Architects and Surveyors Association who, from time to time, publish information sheets on materials and techniques used in relation to cathedrals and churches as well as invaluable guidelines for quinquennial inspections. English Heritage also issue an informative quarterly bulletin and publish books and reports relating to conservation – their current publications relate to subjects as diverse as 'Victorian and Edwardian Stained Glass', 'Directory of Public Sources of Grants for the Repair and Conversion of Historic Buildings', and 'A Farewell to Fleet Street', the last named being a historical and architectural study of Fleet Street and the London newspaper industry. Other useful literature is available such as 'How Old is Your House?', 'Care for Old Houses' and 'Change of Use', all by Pamela Cunnington, and lastly the entertaining and important 'Blue Guide to Victorian Architecture in Britain' by Julian Orbach.

On a purely practical note, 'Practical Building Conservation', by John and Nicola Ashurst is to be recommended, being published in five volumes as follows:

Vol 1. *Stone Masonry*
Vol 2. *Terracotta, Brick and Earth*
Vol 3. *Mortars, Plasters and Renders*
Vol 4. *Metals*
Vol 5. *Wood, Glass and Resins*

These volumes cover the cleaning, repair, use and treatment of the above named items in detail and are all published by Gower Technical Press of Aldershot.

English Heritage also offer repair grants for selected buildings. Typical examples are a grant of £418 000 for the maintenance, repair and renovation of the Chatham Historic Dockyard and £250 000 to North Tyneside Metropolitan Borough Council for the repair of Tynemouth Railway Station as well as smaller projects such as £66 557 towards roof repairs at the Church of Ascension, Lavender Hill, London. Such works have generally been well-planned, closely supervised and information on the scope of work done will be readily available to surveyors generally. The advent of the National Lottery and the possibility of grants from this source may be good news for charities struggling to maintain old buildings.

Although some designers and many developers may say that current concern with conservation represents unhealthy romantic nostalgia for the past – and consequently emasculates technological progress in design and construction – our national culture depends on retaining the best of our living history.

Such is the variation in building techniques in different parts of the country that local knowledge plays an essential part in this type of work and all this presents a fascinating area for study for any keen young surveyor interested in historic buildings.

The conservation bandwagon is now well and truly rolling and surveyors are jumping on board. In 1987 the RICS Conservation Group was formed and two years later had attracted 400 members, about half of them specialist building surveyors. One of the aims of the group was to launch a post-graduate diploma in building conservation at the College of Estate Management at Reading. This is now in place so Chartered Surveyors who wish to specialise in the conservation of old buildings have the opportunity to undertake special training.

Not all surveyors will find themselves undertaking a quinquennial survey of a major cathedral or advising on the restoration of a Grade 1 manor house, but on a more mundane level there is still a contribution that we can all make as we undertake our routine work. Valuers and surveyors are front-line professionals who daily come into contact with the nation's building stock. In the process of reporting we can advise and warn about existing defects and also – equally important from a conservation point of view – we can advise about those aspects of the construction on which a little money spent now can save a far greater expenditure later on.

The use of chemicals in conservation

For some time concerns have been expressed about the use of chemicals in buildings as part of the conservation process. Chemicals are used to reduce or prevent rising damp in masonry. Timbers are routinely sprayed with insecticidal and fungicidal fluids to control rot or beetle attack. The operatives using the equipment for chemical injection or spraying and the occupants of buildings being treated may need reassurance.

In 1991 the Health and Safety Executive published a report 'Remedial Timber Treatment in Buildings – a guide to good practice and the safe use of wood preservatives' with recommendations. Surveyors are advised to obtain a copy. In broad terms the Executive recommends that wholesale treatment of timbers be avoided where possible and that defects such as dry rot and furniture beetle be prevented by good design with less reliance placed on chemical treatment. If timbers are dry and well ventilated most timber defects are avoided and furniture beetle is unlikely to be able to survive in timbers with a moisture content of twelve per cent or less. Widespread rot or beetle attack in a building is an indication that the timbers are damp and that there are design faults.

Conclusion

The reader, having reached this point, will by now appreciate something of the range of knowledge and skill needed to undertake Building Surveys and the contribution that Surveyors can make by ensuring an informed property market and long-term conservation of the building stock. Much still needs to be done to inform and advise the general public about the services offered by Surveyors and the importance of taking expert professional advice. It is hoped that this volume will contribute towards improved professional standards and as public awareness increases the demands for Surveyors' services will continue to grow.

Bibliography

'Shelling of Plaster Finishing Coats', Building Research Establishment (BRE) Information Sheet, TIL 14 1971

'Settlement of a Brick Dwellinghouse on Heavy Clay, 1951–1973', W. H. Ward, BRE, CP 37/74, February 1974

'High-alumina Cement Concrete in Buildings', BRE, CP 34/75, April 1975

'A Simple Pull-out Test to Assess the Strength of *in situ* Concrete', BRE, CP 25/77, June 1977

'Field Measurement of the Sound Insulation of Plastered Solid Brick Walls', E. C. Sewell and R. S. Alphey, BRE, CP 37/78, March 1978

'Foundations for Low-rise Buildings', BRE, CP 61/78, September 1978

'The Performance of Cavity-wall Ties', BRE Information Paper, IP 4/81, April 1981

'House Longhorn Beetle Survey', R. Geraldine Lea, BRE Information Paper, IP 12/82, July 1982

'Inspection and Maintenance of Flat and Low-pitched Timber Roofs', W. T. Hide, BRE Information Paper, IP 15/82, August 1982

'Considerations in the Design of Timber Flat Roofs', I. S. McIntyre and D. P. Birch, BRE Information Paper, IP 19/82, Septemer 1982

'Assessment of Damage in Low-rise Buildings', *BRE Digest* 251, July 1981 (from HMSO)

'The Durability of Steel and Concrete', Parts 1, 2 and 3. *BRE Digests* 263, 264 and 265, July, August and September 1982 (from HMSO)

'Specifications for Roofing Felt', British Standards Institution (BSI), BS 747:1977

'Metal Ties for Cavity-wall Construction', BSI, BS 1243:1978

'Guide to the Choice, Use and Application of Wood Preservatives', BSI, BS 1282:1959 (revised 1975)

'Preservative Treatment for Constructional Timber', BSI, BS 5268, Part 5:1977 (supersedes CP 98)

'Code of Practice for Trees in Relation to Construction', BSI, BS 5837:1980

'Foundations and Sub-structures for non-Industrial Buildings of not more than Four Storeys', BSI, CP 101:1972

'Structural Recommendations for Load-bearing Walls', BSI CP 111:1970

'The Structural Use of Timber', BSI, CP 112, Part 2:1971

'Roof Coverings: Mastic Asphalt', BSI, CP 144, Part 4:1970

'Small Sewage Treatment Works', BSI, CP 302:1972 1st revision. Revised as 'Code of Practice for Small Sewage Treatment Works and Cess-pools', BS 6297:1983

'Foundations', BSI, CP 2004:1972

Building Regulations, and Building Regulations (Amendments), Her Majesty's Stationery Office (HMSO)

'Common Defects in Buildings, H. J. Eldridge, Property Services Agency (PSA), Department of the Environment (DoE), HMSO, 1976

Small Claims in the County Court, HMSO

Quality in Traditional Housing (series), R. B. Bonshor and H. W. Harrison, HMSO, 1982: Vol 1: *Investigation into Faults and Their Avoidance*: Vol 2: *Aid to Design*: Vol 3: *Aid to Site Inspection*

Registered House-Builders' Site Manual, National House-Building Council (NHBC), 1974

Registered House-Builders' Handbook, 1974, and Practice Notes, NHBC

'Timber-framed Buildings', NHBC Practice Note 5 (1982), rewritten and reissued November 1982

'Suspended-floor Construction for Dwellings on Deep-fill Sites', NHBC Practice Note 6, August 1973 revised April 1974

'Structural Surveys of Residential Property', Royal Institution of Chartered Surveyors (RICS), Practice Note, May 1981 (from Surveyors' Technical Services (STS))

Buying a House? RICS 1981 (from STS)

Guide to House Rebuilding Costs for Insurance Valuation (annual), RICS Building Cost Information Service (BCIS) with British Insurance Association

'Code of Measuring Practice', jointly RICS/ISVA (Incorporated Society of Valuers and Auctioneers), 1979

Manual of Timber-frame Housing (A Simplified Method), (National Building Agency) NBA/TRDA (Timber Research and Development Association), The Construction Press, 1980

Timber-frame Housing Design Guide, Section 9, 'Structural Recommendations', TRDA, 1979

Trees in Britain, Europe and North America, Roger Phillips, Pan Books Ltd, 1978

'Building Defects', *The Chartered Surveyor* Supplement, April 1981

'Regulations for Domestic Equipment', Institution of Electrical Engineers (IEE)

Mitchell's Building Construction (series), B. T. Batsford Ltd, Vol 1: *Structure and Fabric*, Part 1, 2nd edn 1979: Vol 2: *Structure and Fabric*, Part 2, 2nd edn 1977: Vol 3: *Materials*, 2nd edn 1978: Vol 4: *Components*, 2nd edn 1979

Building Techniques (series – Science Paperbacks), Chapman & Hall, Vol 1: *Structure*, 2nd edn 1979: Vol 2: *Services*, 3rd edn 1980

Principles of Element Design, Peter Rich, George Godwin Ltd, 1977

Design Failures in Buildings (series), First Series 1972, Second Series 1974, George Godwin Ltd

The Law and Practice of Arbitrations, John Parris, George Godwin Ltd, 1974

Builders' Detail Sheets, Series 2, Northwood Publications, 1977

'A. J. Guide to Structural Surveys', Clive Richardson, *Architects Journal* 1985

Premises Management, Croner Publications Ltd, 1987

Maintenance Management, Chartered Institute of Building 1989

The RICS Property Doctor Book, RICS & BBC Television

Building Maintenance Management, Reginald Lee, Crosby Lockwood Staples, 1976

Shaw's Commentary on the Building Regulations, Shaw & Sons Ltd, 1977
How to Study Architecture, Charles H. Caffin, New York Tudor Publishing Co, 1937
Rebuild, R. Derricott and S. S. Chissick, John Wiley & Sons Ltd, 1982
Quality and Total Cost in Buildings and Services Design, The Construction Press, 1976
Architects' Standard Catalogues (annual) Standard Catalogue Information Services Ltd
Spon's Architects' and Builders' Price Book (annual), E. & F. N. Spon
The Enemies of Timber, The Cuprinol Preservation Centre, 1971
Alteration or Conversion of Houses, 4th edn, J. F. Garner, Oyez Press Ltd, 1975
Nothing But The Truth, John Watson, The Estates Gazette Ltd, 1971
Real Property Law, Hugh H. V. Forbes, The Estates Gazette Ltd, 1962
Remedial Timber Treatment in Buildings – a guide to good practice and the safe use of wood preservatives, HMSO Publications, 1991

Related Books from Butterworth-Heinemann, Architectural Press and Laxton's

New Metric Handbook, Architectural Press
Basic Metric Surveying, 4th edn, W. S. Whyte and R. E. Paul, 1977
A History of Architecture, 4th eds, Prof. Banister Fletcher, FRIBA, and Banister F. Fletcher, ARIBA, B. T. Batsford Ltd, 1901. 20th edn, edited by Dan Cruickshank, Architectural Press, 1996
Performance of Materials in Buildings, L. Addleson and C. Rice, 1995 (paperback edn)
Building Failures, 3rd edn. L. Addleson, 1992
Dampness in Buildings, 2nd edn, T. A. Oxley and E. G. Gobert, 1994
The Conservation of Historic Buildings, Revised edn, Sir Bernard Feilden, 1994
The Conservation of Building and Decorative Stone, J. Ashurst and F. G. Dimes, 1990
Laxton's Building Price Book (annual)
Resealing of Buildings, R. Woolman and A. Hutchinson, 1994
Building Terminology, P. Brett, 1993
Hutchins' Priced Schedules (annual)
Building Maintenance and Preservation, 2nd edn, E. Mills, 1994
Care and Conservation of Georgian Houses, 4th edn, 1995
Masonry Walls, K. Thomas, 1995
Builders Estimating Data, D. Cross, 1990
Building and Land Management Law for Students, A. Galbraith and M. Stockdale, 1993
Building Construction Handbook, 2nd edn, R Chudley, 1995
Construction Industry of Great Britain, R. Harvey and A. Ashworth, 1993
Construction Materials Reference Book, D. Doran, 1994
Construction Materials Pocket Book, D. Doran, 1995
Remedial Treatment of Buildings, 2nd edn, B. Richardson, 1995
Contract Practice for Surveyors, 3rd edn, J. Ramus and S. Birchall, 1996
Envelope Design for Buildings, W. Allen, 1997

Acts of Parliament and other Regulations

Agricultural Holdings Act 1948
Building Act 1984
Building Regulations 1985
Defective Premises Act 1972
Factories Acts
Fire Precautions Act 1971
Fire Precautions Act 1971 (Modifications) Regulations 1976
Health and Safety at Work Act 1974
Historic Buildings and Ancient Monuments Act 1953
Homes Insulation Scheme 1982
Housing Acts 1957, 1961, 1974 and 1980
Housing Finance Act 1972
Landlord and Tenant Acts 1927, 1954 and 1985
Law of Property Act 1925
Leasehold Property (Repairs) Act 1938
Leasehold Reform Act 1967
Limitation Acts 1939 and 1980
Local Authorities (Historic Buildings) Act 1962
London Building Acts and London Building Acts (Amendment) Act 1939
London Building (Construction) Bye Laws 1979
Offices, Shops and Railway Premises Act 1963
Public Health Act 1939
Rent Acts
Supply of Goods and Services Act 1982
Town and Country Amenities Act 1974
Town and Country Planning Act 1971
Unfair Contract Terms Act 1977

Court cases and page reference

Acrecrest Ltd *v* W. S. Hattrell & Partners (a firm) [1 *All ER* 17, [1982] 3 *WLR* 1076   219
Anns *v* Merton London Borough Council [1977] 2 *All ER* 492, [1978] AC 728   215
*Balcomb *v* Wards Construction (Medway}) Ltd (1981) 259 *EG* 765   220
Batty *v* Metropolitan Property Realisations [1978] 2 *All ER* 445, [1978] QB 554   219
*Bolam *v* Friern Hospital Management Committee [1957] 2 *All ER* 118, [1957] 1 *WLR* 582   217
Bunclark *v* Hertfordshire County Council (1977) 243 *EG* 455   30
Cross and Another *v* David Martin & Mortimer 18 November 1988   129
*DoE *v*Thomas Bates and Son 26 July 1990   215
*Donoghue (or McAlister) *v* Stevenson [1932] *All ER* Rep 1, [1932] AC 562   216
Drummond *v* S & U Stores (1980) 258 *EG* 1293   174
*Dutton *v* Bognor Regis Urban District Council [1972] 1 *All ER* 462, [1972] 1 QB 373   215

*Eames London Estates Ltd *v* North Hertfordshire District Council
(1981) 259 *EG* 491   218
Finchbourne Ltd *v* Rodrigues [1976] 3 *All ER* 581   144
*Fisher *v* Knowles (1982) 262 *EG* 1083   223
*Fryer *v* Bunney (1982) 263 *EG* 158   222
Hancock *v* B. W. Brazier (Anerley) Ltd [1966] 2 *All ER* 901, [1966]
1 *WLR* 1317   148
Hatfield *v* Moss [1988] 40 *EG*112   231
*Hedley Byrne & Co Ltd *v* Heller & Partners LTd [1963] 2 *All ER* 575,
[1964] AC 465   214
*Leigh *v* Unsworth (1972) 230 *EG* 501   213
*London & South of England Building Society *v* Stone (1981) 261
EG 463   221
*McGuirk *v* Homes (Basildon) Ltd and French Kier holdings Ltd (*Estates
Times* 29 January 1982)   221
Mather *v* Barclays Bank Plc [1987] 2 *EGLR* 254   232
*Morgan *v* Perry [1973] 229 *EG* 1737   212
*Murphy *v* Brentwood DC 26 July 1990   215
Murray *v* Birmingham City Council [1987] 283 *EG* 962   234
Norwich Union Life Assurance Society *v* British Railways Board [1987] 283
EG 846   233
*Perry *v* Sidney Phillips & Son (a firm) [1982] 1 *All ER* 1005, [1982] 3
All ER 705, [1982] 1 *WLR* 1297   217
Phillips *v* Price [1959] Ch.181   235
Pirelli General Cable Works Ltd *v* Oscar Faber & Partners [1983] 1 *All ER*
65, [1983] 2 *WLR* 6   214
Rich Investments Ltd *v* Cumgate Litho Ltd [1988] *EGCS* 132   234
*Rona *v* Pearce [1953] 162 *EG* 380   217
Sidnell *v* Wilson 1966   234
*Stevenson *v* Nationwide Building Society [1984] 25 July 1984   225
Straudley Investments Ltd *v* Barpress Ltd [1987] 282 *EG* 1224   231
*Treml *v* Ernest W. Gibson & Partners [1984] 27 June 1984   224
*Watts and Another *v* Morrow [1991] 14 *EG* 111   226
*Yianni *v* Edwin Evans & Sons (a firm) [1981] 3 *All ER* 592 [1981] 3 *WLR*
843   215

*Cases discussed in Chapter 19

Useful addresses

National House-Building Council (NHBC)
Chiltern Avenue, Amersham, Bucks HP6 5AP
Building Research Establishment (BRE),
Building Research Advisory Service (BRAS)
Garston, Watford, Herts WD2 7JR
Department of the Environment (DoE),

Property Services Agency (PSA)
2 Marsham Street, London SW1P 3EB
Royal Institution of Chartered Surveyors (RICS),
Surveyors' Technical Services (STS – publishers for RICS)
12 Great George Street, London SW1P 3AD
Building Cost Information Service (BCIS) (RICS),
Building Maintenance Cost Information Service (BMCIS) (RICS)
85–87 Clarence Street, Kingston-upon-Thames, Surrey KT1 1RB
Royal Institute of British Architects (RIBA)
66 Portland Place, London W1N 4AD
Institution of Electrical Engineers (IEE)
Savoy Place, London WC2R 0BL
Incorporated Society of Valuers and Auctioneers (ISVA)
3 Cadogan Gate, London SW1X 0AS
British Standards Institution (BSI)
Head Office: 2 Park Street, London W1A 2BS
Sales: 101 Pentonville Road, London N1 9ND
Her Majesty's Stationery Office (HMSO)
Government Bookshop, 49 High Holborn, London WC1V 6HB
Society for the Protection of Ancient Buildings
55 Great Ormond Street, London WC1N 3JA
Timber Research and Development Association (TRDA)
Chiltern House, Stocking Lane, Hughenden Valley, High Wycombe,
Bucks HP 14 4ND
Ordnance Survey
Romsey Road, Maybush, Southampton, Hants SO9 4DH
Institute of Geological Sciences
Exhibition Road, London SW7 2DE
British Insurance Association
Aldermary House, Queen Street, London EC4N 1TU
Brick Development Association
Winkfield, Windsor, Berks SL9 2DP
Cement & Concrete Association
52 Grosvenor Gardens, London SW1W 0AQ
Institution of Structural Engineers
11 Upper Belgrave Street, London SW1X 8BH
British Wood Preserving Association
Premier House, Southampton Row, London, WC1B 5AL
Chartered Institute of Building
'Englemere', Kings Ride, Ascot, Berks

Index